Solar Energy in Agriculture

Solar Energy in Agriculture
Principles and Applications

Editors
Priyabrata Santra
Ranjay Kumar Singh
Surendra Poonia
Dilip Jain

ICAR-Central Arid Zone Research Institute
Near Industrial Training Institute
CAZRI Road, Jodhpur
Rajasthan-342003, India

NEW INDIA PUBLISHING AGENCY
New Delhi – 110 034

First published 2021
by CRC Press
2 Park Square, Milton Park, Abingdon, Oxon, OX14 4RN

and by CRC Press
6000 Broken Sound Parkway NW, Suite 300, Boca Raton, FL 33487-2742

CRC Press is an imprint of Informa UK Limited

British Library Cataloguing-in-Publication Data
A catalogue record for this book is available from the British Library

Library of Congress Cataloging-in-Publication Data
A catalog record has been requested

ISBN: 978-0-367-72659-1 (hbk)

Preface

Solar energy is the clean source of energy and it is abundantly available in nature. Thus, it has a huge potential to supplement the fast depleting fossil fuel, which is otherwise emitting greenhouse gasses (GHG) in atmosphere. Therefore, solar energy is considered as a good option for mitigating climate change effect in future. Broadly solar energy can be utilized in two ways. One is to convert solar energy to electrical energy through photovoltaic process, which is known as photovoltaic (PV) technology. Another is to convert solar energy to heat by trapping the incident solar energy in to a closed enclosure, which is also known as solar thermal technology. Till few years ago, solar thermal technology was preferred over solar PV technology because the cost of installation for PV system was quite high. However, the situation is just opposite now. Recently the PV technology has been preferred because the cost of PV systems has been reduced drastically during last few years. Nowadays, PV modules are available in Indian market at a cost of Rs 30-35 W_p^{-1}. Throughout the world, increased attention has been vested upon solar energy installations. National solar mission has also been in progress in India with a target of 100 GW grid-tied solar PV installations by the end of 2022. Similarly, off-grid PV generation target is 2 GW, which includes solar PV pumping system. Apart from PV generation, there is target of installing 20 million m^2 solar thermal collectors e.g. solar drier, solar cooker, solar water heater etc. Simultaneously, there is also target of 60 GW wind energy generation, 10 GW biomass power generations and 5 GW from other renewables, adding the total renewable energy target of India to 175 GW by 2022. In this context, the book on "Solar energy in agriculture: principles and applications" is appropriate and timely.

Agriculture sector consumes about 7-8% of total energy consumption of India. Pumping of irrigation water, use of heavy machineries for different farm operations, processing and value addition of farm produces etc. are major activities by which energy is consumed in agriculture sector. With the advancement of food production system from agrarian to a futuristic technology-driven system, there has been rapid increase in energy use in agriculture. It has been expected that energy use in agriculture in Indian needs to be increased from its present value 1.6 kW ha^{-1} to 2.5 kW ha^{-1} to meet the production target of next 20 years. In this context, we need to harness and use more renewable forms of energy, especially solar energy that is plentiful on most part of the country.

Agriculture sector has great scope in meeting the solar energy installation targets at different parts of the world. This can be achieved through major two ways. First is the replacement of fossil fuel based farm operations with solar energy based devices and implements. Second is the contribution in renewable energy generation from agriculture sector. The first approach includes replacing diesel operated or grid-tied electric pumps with solar PV pumping system, use of solar devices for processing and value addition of foods, increasing use of solar PV driven tools and implements etc. The second approach is through contribution in renewable energy generation may be achieved through either cogeneration of food and energy using agri-voltaic and solar-wind hybrid system or utilizing biomass and agro-wastes for energy generation.

In this book we focus to discuss all these possible options of solar energy use and generation in agriculture sector. However, before to it, basic fundamentals of solar energy resources and technologies are discussed in detail. Overall, the book contains 23 chapters. Out of these, first two chapters focus on solar energy use pattern in agriculture sector in India at present time along with future scopes. The next eight chapters (Chapter Nos 3-10) give a basic knowledge on fundamental principles of solar photovoltaic and thermal technologies. Last 13 chapters (Chapter Nos 11-23) presents the applications of solar thermal and photovoltaic technology in different farm operations and postharvest processing in agriculture sector. We hope the book will cater the needs for students, researchers, various stakeholders, entrepreneurs etc by providing valuable information on solar energy and its applications specifically focusing on agriculture.

Editors

Contents

1

Energy Requirements in Agriculture and Renewable Energy Options for Future

Priyabrata Santra and O.P. Yadav

ICAR-Central Arid Zone Research Institute, Jodhpur, Rajasthan, India

Introduction

In order to keep pace with the development there is rise in energy use but it has adverse effects on climate due to greenhouse gas emissions from burning of fast depleting fossil fuels. In this context, we need to harness and use more and more renewable forms of energy, especially solar energy that is plentiful on most part of India. Also, at several locations harnessing wind power and utilizing biomass could be effective alternatives. Solar based devices may also work in an integrated manner with small wind turbines as hybrid devices. At present, about 20% of the country's installed electricity generation capacity is contributed by renewable sources e.g. wind, solar, bioenergy, hydro etc., which is about 71.325 Giga Watt (GW) as on 30th June, 2018. In agricultural sector, energy is directly used for pumping irrigation water, operating different mechanized farm implements/tools and processing of foods. Share of agricultural sector in total energy consumption is about 7-8% and further increase in energy use from its present value of 1.6 kW ha^{-1} to 2.5 kW ha^{-1} is expected to meet the production target of next 20 years.

Availability of solar irradiance in India

The arid and semi-arid part of the country receives much more radiation as compared to the rest of the country. The average irradiance on horizontal surface in India is 5.6 kWh m^{-2} day^{-1} and at Jodhpur 6.11 kWh m^{-2} day^{-1}. The solar resource map of India shows that western India receives maximum amount of solar irradiation whereas major portion of India (~140 million ha) is receiving solar irradiation of 5-5.5 kWh m^{-2} day^{-1} (Fig. 1.1). The solar resource map along with grid wise solar radiation data can also be downloaded from http://mnre.gov.in/sec/solar-assmnt.htm. The cold arid region of the country located at Leh and Ladakh receives highest amount of radiation, which is about 7-7.5 kWh m^{-2} day^{-1}. At Jodhpur, maximum amount of radiation is received during the month of April (7.17 kWh m^{-2} day^{-1}), whereas the minimum amount of radiation is received during the month of December (5.12 kWh m^{-2} day^{-1}). In total, 6390 kWh of solar energy is available during a year at Jodhpur. Moreover, most of the days in a year at Jodhpur are cloud free which has been measured and reported in several literatures as 300 days clear sunny days in a year. Available solar irradiation and utilizable energy for any location in India can also be viewed from http://pvwatts.nrel.gov/ or http://mnre.gov.in/sec/solar-assmnt.htm.

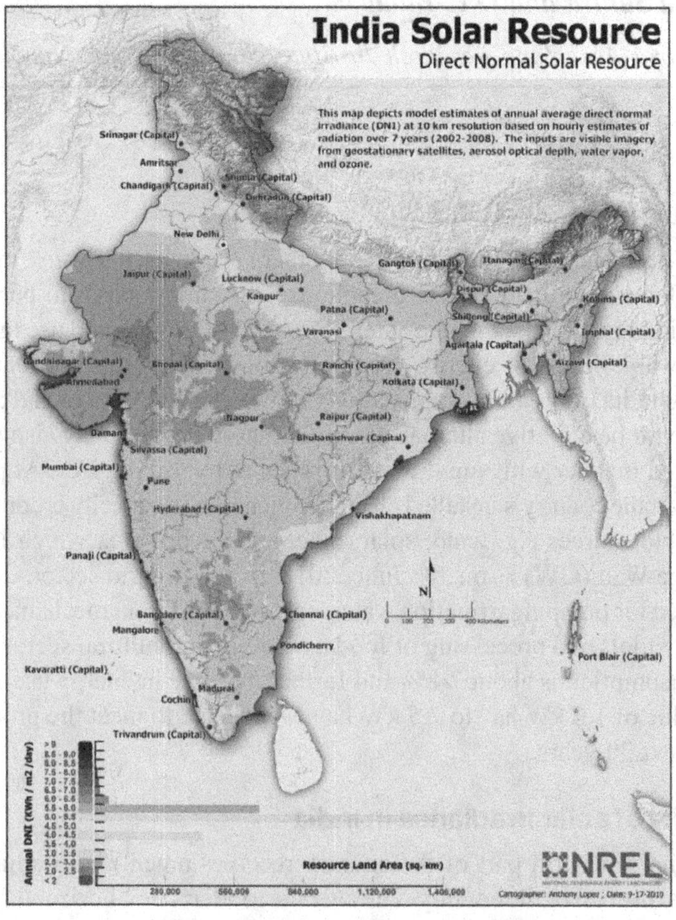

Fig. 1.1: Solar resource map of India
Source: Solar Energy Centre, GoI, India

Renewable energy scenarios in World vis-a-vis India

At present, renewable energy share to world's global electricity production is about 26.5% (by the end of 2017), out of which 16.4% is contributed by hydropower, 5.6% by wind energy, 2.2% by biomass-power and 1.9% by solar PV (Renewable Energy Network for 21st Century, REN21). Cumulative renewable installed capacity in the world is 2195 GW. Annual growth rate of cumulative renewable energy installed capacity in 2017 was about 8.8%, whereas the annual capacity addition grew by 7% in 2017 as compared to 2016. India ranks 5th in the world in total renewable energy installed capacity while China tops the list followed by USA and Brazil. In China, wind energy and hydropower installations are the major contributors to renewables whereas in USA, geothermal energy and in Germany, solar PV is the dominant contributor. India ranks 4th in the world in total wind energy installation after China, USA and Germany, whereas it is 10th in world among solar PV installation. Globally, 15% of the world population has no access of electricity. India today is home to one-sixth of the world's population, but accounts for only 6% of global energy use and one in five of the population – 240 million people – still lacks access to electricity (World energy council, 2015). Therefore, much effort is needed in India to fulfill the future energy demand and specifically through renewable energy sources.

At present about 20% of energy generation in India is met through renewable sources e.g. wind, solar, biomass etc. whereas coal is till the main source contributing about 57% of total generation. During last few years, a great stride has been made to install solar PV plants, wind turbine, hydropower, biogass e.g. renewable installed cumulative capacity has been increased from 24914 MW in 2011-12 to 69022 by the end of 2017-18 with an annual growth rate of 17.8% (Fig. 1.2).

By the end of March 2018, wind energy installation shares the maximum 34,046 MW (49.3%) whereas solar PV installation shares 21,651 MW (31.4%). Rajasthan and Gujarat share ~22.9% of the total solar power installed capacity in the country, whereas these two states shares 48.3% of total wind installed capacity. Tamilnadu and Maharastra dominate the total wind installation in our country by sharing 33.4% of total installed capacity by the end of 2017-18. Simultaneously 60,000 MW of wind energy is targeted to achieve by 2022. Among these national targets, Rajasthan and Gujarat shares the maximum (8600 and 8800 MW, respectively).

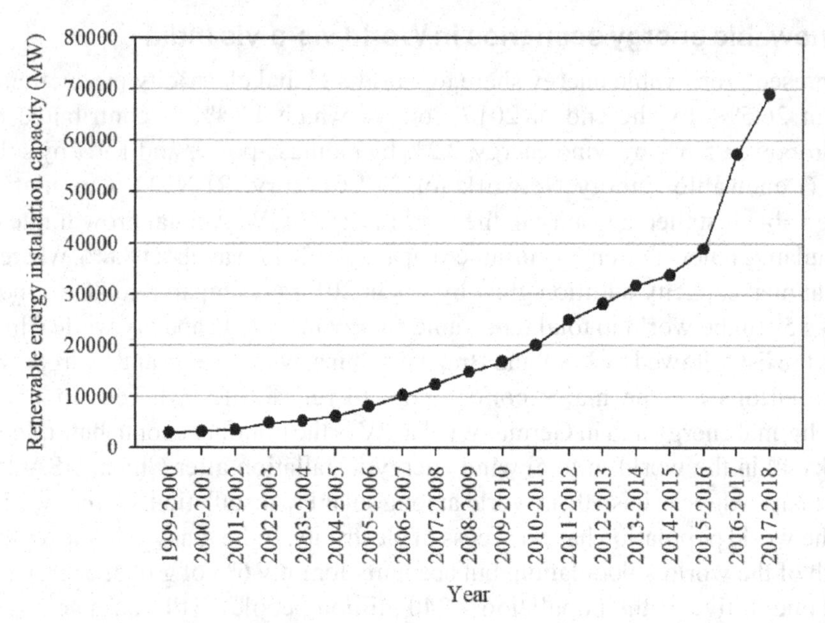

Fig. 1.2: Renewable energy installation capacity in India

National solar mission

The National Solar Mission (NSM) has been in operation since 2010 with the following targets in three phases (Table 1.1). The target has been revised in 2015 to a total grid connected solar power generation of 1,00,000 MW comprising 40,000 MW roof top generation and 60,000 MW grid connected solar power plants (Resolution of MNRE, Govt of India, No. 30/80/2014-15/NSM dated 1st July 2015).

Table 1.1: National solar mission targets

Sr.No.	Application segment	Target for Phase I (2010-13)	Target for Phase II (2013-17)	Target for Phase III (2017-22)
1.	Grid connected solar power generation	1,100 MW	4,000 MW	1,00,000 MW*
2.	Off-grid solar applications (includes solar PV pump)	200 MW	1,000 MW	2,000 MW
3.	Solar thermal collectors	7 million sq. m.	15 million sq. m.	20 million sq. m.
4.	Solar lighting systems	5 million	10 million	20 million

*The revised target (Source: Ministry of Renewable Energy Sources, Govt. of India)

In Rajasthan several initiatives have been taken to support the growth targets of renewable energy in India by Rajasthan Renewable Energy Corporation Limited (RRECL) (www.rrecl.com). Total wind energy installed capacity in

Rajasthan up to December 2018 is 4299.72 MW whereas total grid connected solar PV installed capacity is 3130.99 MW. Solar power park has been developed at Badhla, Phalodi with an area of 1360 ha and expected generation capacity of about 600-700 MW. MoU has been signed between RRECL and Suzlon to develop solar wind hybrid project of 1500 MW capacity (http://www.rrecl.com/PDF/MOUSUZLON.pdf). Similarly, horticulture department of Rajasthan Govt. has been constantly devoted their efforts to install solar PV pumping system (3 HP and 5 HP capcity) in farmer's field for irrigation purpose (http://horticulture.rajasthan.gov.in/ Content Detail. aspx? pagename= National_ Solar_ Mission).

Renewable energy options in agriculture

Considering the potential of solar energy, few avenues of its utilization in agriculture is given below:

(i) **Solar PV operated water lifting / pumping system**: Water is the primary source of life for mankind and one of the most basic necessities for crop production. The demand for water to irrigate the crops is increasing. For sustainable production from agricultural farms, irrigating the crops at right stages is highly important. Even in rainfed situation, lifesaving irrigation during long dry spell has also been found beneficial for crop survival and to obtain the targeted yield. Pressurized irrigation systems e.g. drippers, sprinklers etc are of great importance in 'crop per drop' mission, however, ensured power supply is essential to operate these systems. Solar PV pumping systems may be quite helpful to operate the pressurized irrigation system. Specifically, solar pumps may be useful as water lifting devices in irrigation canals and also to evenly distribute water in command areas and thus will reduce the wastage of water. At present, about 16 million electric pumps and 7 million diesel pumps are in operations in the country for irrigation purpose; however they are highly energy intensive and therefore if replaced with solar pumps may greatly contribute in country's energy security. Till December 2017, 38,687 pumps have been installed in the country, which are mostly of 2 or 3 HP pumping system, which has been recently appended with 5 HP pumping system. These solar pumps have the capacity to withdraw water from a depth of about 75 m and therefore may be beneficial in those areas where groundwater is not deeper than it. Moreover, solar pumps are directly operated by solar irradiance and therefore diurnal and seasonal variations of it play a key role in implementation of solar PV pumps in a place. In arid and semi-arid regions of the country, clear sky condition with average irradiance of 5-6 kWh m^{-2} day^{-1} is available for 300 days in a year and thus solar pumps can be operated for about 6 hours a day and most of the period in a year.

(ii) **Agri-voltaic or Solar farming:** Agri-voltaic land utilization system or solar farming is proposed to either ascertain a portion of land for erection of PV modules in a famer's field or introducing crop cultivation in the same piece of land where PV panels are erected for electricity generation purpose. By adopting such system, the risk of loss due to crop failure during aberrant weather events may be marginalized in farm scale and may prove to be an effective drought proofing strategy. PV panels are placed in a solar power plant for electricity generation conventionally in long rows with sizeable areas left blank by default to avoid shading of one row by another. These inter- panel areas and below -panel areas can be effectively utilized for growing such crops that can tolerate certain amount of shade for different durations of the day. In arid zone this small amount of shading actually serves as a boon by stopping evaporation of water as well as reducing transpiration losses. Secondly, all solar PV plants in arid zone have a problem of deposition of a fine film of wind born sand on panels requiring a water spray to clean it. This water washes down the panel into the soil. Thus, there is an availability of partially shaded space and recurrently available washout water, which can both be harnessed for growing crops. Ideally, crops for these sites should be such that it is not taller than 50-70 cm, preferably perennial, spreading, and do not interfere in any way with the functional efficiency of solar power plant.

(iii) **Solar based processing of agricultural produces**: Processing of agricultural produces like drying, cleaning, grading, winnowing etc. is important for value addition. There are already well established solar thermal and PV based devices available commercially and many specific technologies have been developed by different institutes for farmers and villagers. For example, inclined solar drier have been found quite useful to dry different agricultural produces along with maintenance of quality of the produce. Animal feed solar cooker have also been found to augment the milk production from cattle by providing them quality boiled feed. Solar water heater also has great potential in different processing stages. Solar PV winnower cum drier helps in cleaning of agricultural produces and also helps in preparing dried products. Solar PV operated duster helps in applying agricultural chemicals in agricultural field to protect crops from pests and diseases. Passive cool chamber may be useful for on-farm short term storage and preservation of fruits and vegetables.

(iv) **Solar PV hybrid devices:** Small wind aero-generators in hybrid mode with solar panels are useful for off grid renewable energy based electricity generation. These devices are quite useful considering their 24 hour generation potential. Solar PV panel is capable of generating electricity during day time and clear sky conditions, whereas small wind turbine of Savonious design at low heights is capable of harnessing wind energy during both day and night and even on cloudy days. In agricultural farms, installation of such hybrid devices along farm boundary not only will protect crops but also will generate electrical energy that can be used in different farm operations.

Co-generation of renewable energy

Till now, solar PV and wind turbine has been installed separately on land. If we see the wind resource map and solar atlas of India, there are few zones of the country where both wind and solar resources are available in plenty (Fig. 1.3). There are zones in Rajasthan and Gujarat where wind power density at 50 m height above ground level (agl) is > 200-250 W m^{-2} and simultaneously solar resources are also high (5.6-6.6 kWh m^{-2} day^{-1}). A portion of Jammu and Kashmir lying at high altitude has also high wind and solar resources, however due to difficult terrain and topography it is not always feasible to install solar and wind turbine.

The requirement of land for both wind and solar installation is a key issue and generally non-productive lands are being used for this purpose. Land requirement for wind and solar installation are about 0.5 ha MW^{-1} and 2 ha MW^{-1}, respectively. Horizontal axis wind turbines are generally installed at 50 m or greater height agl whereas solar PV plants are installed on ground surface therefore both wind and solar can be installed together in a single piece of land. Such hybrid installation will be able to generate 2.5 MW renewable energy per ha of land (wind: 2 MW ha^{-1} and solar: 0.5 MW ha^{-1}). Another advantage of such hybrid installation is the use a single grid to distribution of generated energy which will further help in reduction of price per unit energy. Apart from energy generation, lands may also be utilized to grow crops or other economic plants since a sizeable portion of land is left blank between two rows of PV arrays in the solar PV installation. It has been observed that 6-12 m inter-row spaces has been kept fallow and therefore a minimum of 1/3rd portion of total land may be available for growing crops and plants. Thus in 1 ha of solar PV plant, 0.67 ha land may be utilized for agriculture purpose. Another advantage of such hybrid system is the possibility of harvesting rainwater from top surface of panels, which may be recycled for maintenance of PV panels and irrigating the crops. It has been estimated that about 1000 m^3 of water may be harvested from 1 ha solar PV plant considering an average annual rainfall of 285 mm and runoff coefficient of 0.8 for Jodhpur condition. Therefore, three-in-one land utilization option for generating electricity and producing food may be a most viable future option for energy and food security of our country.

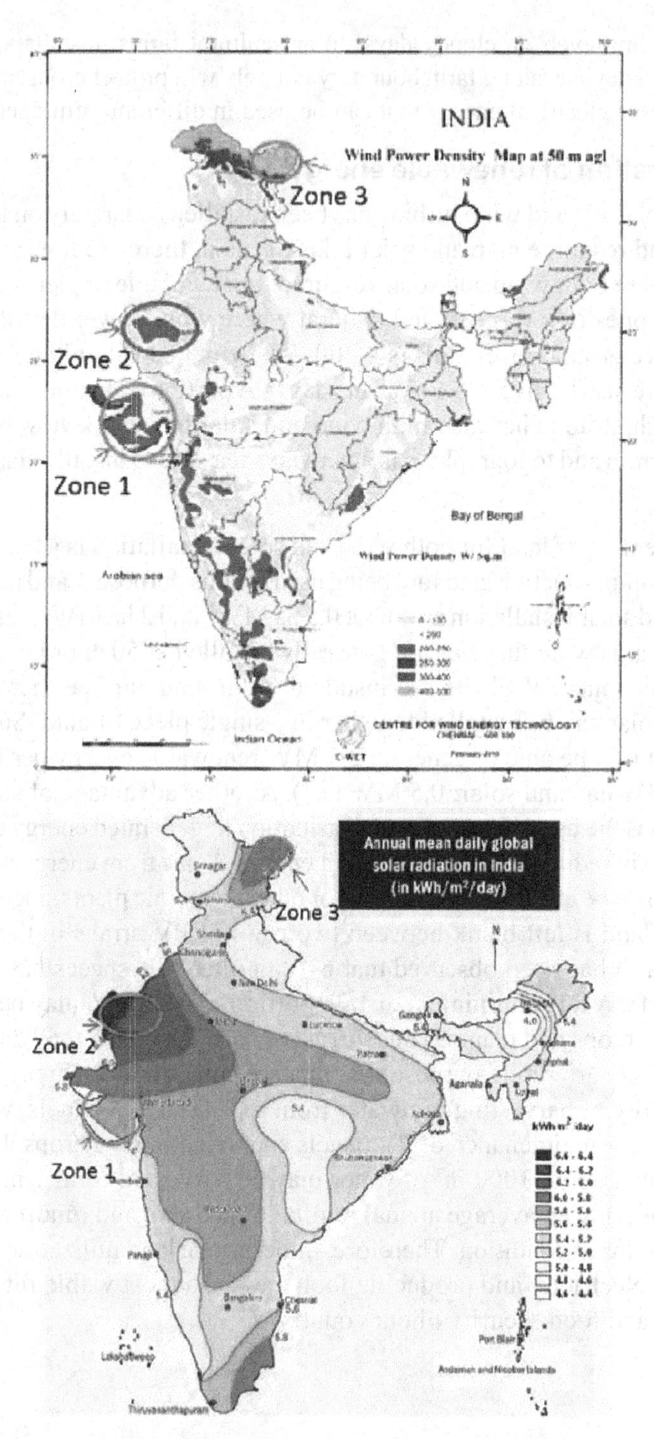

Fig. 1.3: Solar and wind resources in India

Summary

In agricultural production system energy is a key input and is conventionally met by animal drawn power or manual operations which have recently been upgraded by mechanization process. Nowadays, several farm operations are done through various types of machine driven tools and implements e.g. (i) tractor driven harrow, intercultural operation, harvesting, threshing etc; (ii) Diesel or electricity operated pumps for irrigation; (iii) post-harvest processing machines etc. Most of these modern farm tools and implements use either diesel or fossil fuel based energy for its operation. Since, renewable energy in the form of solar and wind energy are abundantly available in India and with advancement of technologies in recent times; there is huge potential to replace these fossil fuel based energy sources. In this paper, future scope of renewable energy in agriculture sector is discussed. The concept of solar farming or agri-voltaic has a special role to eliminate the problem of land utilization for renewable energy installation. Even the concept of hybrid generation of wind and solar energy along with food production may plays a major role in future agriculture in India.

2

A Brief Overview of Solar Energy Initiatives in India: Past, Present and Future

P.C. Pande

Former Principle Scientist, ICAR- Central Arid Zone Research Institute Jodhpur, Rajasthan, India

Introduction

The development of any region is reflected in its quantum of energy consumption. With a view to keeping the pace we have to grow our energy resources at a rate that commensurate to sustainable development. There is already shortage of electricity, reflected in power cuts notwithstanding enhancement of electricity production of several folds in the country. The problem is more severe in context with rural areas in India where some 70% of population live and have agriculture as the main occupation. Although statistically few thousand villages are yet to be electrified, the availability of regular supply in far off places has been a problem and the farmers are unable to derive benefits of electricity. The fast depleting kerosene is used for lighting and diesel oil for running agricultural machinery including pumps. In addition people burn firewood, agricultural waste and cow dung cake for cooking food causing irreparable damage to the eco-system. According to Lester Brown there will be plenty of food to eat but no fuel to cook. The burning of cow dung deprives farmers the use of potential source of organic manure. Further, due to shortage of energy resources, farmers are unable to process the agricultural products to enable them to accrue more benefits. The situation is still worse in arid region where bio mass is scarce and there is no hydro electricity. In addition, farmers are unable to generate additional income due to lack of energy resources to run appropriate device in cottage industries. In this context, solar energy utilization has a tremendous potential for providing needful energy to the farmers for domestic, agricultural and allied applications leading to a sustainable development.

India occupies better position regarding solar energy potential. During winter from November to February most of the Indian stations receive 4.0 to 6.3 $kWhm^{-2}day^{-1}$ solar irradiance, while in summer season this value ranges from 5.0 to 7.4 $kWhm^{-2}day^{-1}$. The arid and semi-arid parts of the country receive much more radiation as compared to rest of the country with 6.0 $kWhm^{-2} day^{-1}$ mean annual daily solar radiation having 8.9 average sunshine hours a day at Jodhpur. Jaisalmer receives maximum radiation i.e., 6.27 $kWhm^{-2}day^{-1}$(Mani and Rangrajan, 1982). Considering this Jodhpur, Barmer and Jaisalmer districts are declared as solar enterprises zone suitable for setting up of solar power plants. The solar irradiance available in cold desert region, such as Leh, were observed to be 5.53 $kWh\ m^{-2}day^{-1}$ on horizontal plane and 6.36 $kWh\ m^{-2}day^{-1}$ at a 35 degree south facing tilt indicating an excellent potential of solar energy in high altitude cold desert (Jacobson 2000).

The Sun God has been worshiped since the dawn of civilization in India. Our ancestors knew the importance of solar radiation and have therefore designated sun radiation as the main source of energy for sustenance. It was not only in India, the earlier civilization of Greece, Egypt, Mesopotamia etc. reflected the sun as a source of power in their scriptures, paintings, sculptures. Famous sun temple in Konark is an example of the same. In nature solar energy is used by plants to prepare food through photosynthesis and stored in biomass. Technologically it can be used in the form of thermal energy by using flat plate collectors for low temperature applications or focusing concentrators for high temperatures or alternatively converting part of incoming radiation directly to electricity through photovoltaic cells. There has been considerable research work that has been carried out all over the world by various workers to harness solar energy efficiently and effectively for practical applications. As a result we have been able to successfully use solar appliances to supplement our day to day energy needs. Here we would like to scan briefly the initiatives that were taken in India at different period of time and specifically the efforts made at CAZRI, Jodhpur to put the historic perspective in conjunction with pragmatism.

Historic perspective

In fifties and sixties little attention was given in India on the development of solar technology. Some isolated interest has been there to develop devices for domestic and industrial applications. A paraboloid solar cooker was developed in NPL by Dr. Ghai, which was commercialized through Devi Dayal company. It required tracking and therefore did not pick up. Another hot box type solar cooker was also developed by Dr Ghosh, and later by Dr Parikh. Efforts were made in CBRI Roorkee on the development of indigenous solar collectors for water heating and air heating by Prof. C.L. Gupta and his team. Dr. Gupta later moved to Aurbindo Ashram at Pondicherry and continued his efforts through design of solar concentrators and other practical devices. His contribution in the field of solar thermal energy is immense. His one popular article of 1975 in Science Today is as relevant and contemporary as it was. Defence lab at Jodhpur with Dr. J.P. Gupta (who later joined CAZRI) and team also did some good work for water heating, snow melting, space heating etc. for cold areas like Leh. Dr.S D Gomkale, VVN Kishore, Mahabala, Shah and team at CSMCRI, Bhavnagar were also involved, specifically in the field of distillation and developed large scale solar stills and installed in village Avania. They did some pioneering work on solar pond also. At that time in the field of solar thermal energy H Tabor of Israel was considered as generator of new ideas in the field of solar thermal. It is interesting to note that at this crucial time a full fledged division of Wind Power and Solar energy was created at CAZRI in 1972 with Dr. A. Krishnan on the chair. In all the National Solar/Renewable Energy Conferences, used to be organized by Solar Energy Society of India, the contribution of CAZRI was significant. A coordinated USAID research programme on Solar Energy for Agriculture was initiated in late 70's with CAZRI Jodhpur (Coordinating Centre), PAU Ludhiana, CTAE Udaipur, TNAU Coimbatore, CRRI Cuttack as collaborators. In this connection the contributions of Dr H. P. Garg, K. P. Thanvi, P. C. Pande, N.M. Nahar need mention. Dr R. P. Singh briefly coordinated it. The role of Prof. Mannan, Prof Cheema and Prof. Parampal Singh of PAU Ludhiana in the project is worth mentioning. Later the project was enlarged and the coordinating centre moved to CIAE Bhopal but CAZRI remained as main Solar Energy Centre with Dr. J. P. Gupta as PI. Simultaneously the young team at CAZRI contributed tremendous work on different aspects of solar energy utilization through institutes' research programs.

Again going back to1973, more emphasis was made to utilize this immense source of energy due to that time prevailing oil crisis. A report on Solar Energy –A Promise and Challenge was prepared by Dr. G. D. Sootha of NPL and was finalized in National Committee on Science and Technology. This was considered as action plan for solar energy utilization research and development work in India. A Department of Non Conventional Energy Sources was created in DST. Later a separate ministry MNES was created. Dr. Maheshwar Dayal from BARC was requested to lead it. Dr. G. D. Sootha was given responsibility of Solar Thermal while Dr. J. Gururaja looked after solar photovoltaics. MNES was later renamed as MNRE with incorporation of other renewable energy sources. The contributions of Dr T. C. Tripathi as Advisor and Mr. Deepak Gupta, Secretary were also immense. Now it is merged with Ministry of Energy. One R & D centre was established at Gwal Pahari, Haryana that was devoted to solar thermal applications initially, and is now designated as NSEI. During this period R & D activities enhanced with establishment of Centre for Energy Research at IIT Delhi, with Prof M. S. Sodha as leader, Prof N. K. Bansal, Prof. H. P. Garg (moved from CAZRI) and many academicians like Prof. Malik, Prof. Kaushik, Prof. G. N. Tewari, Prof. T. C. Kandpal etc. provided huge inputs to research, mostly related to modeling and developing mathematical tools and producing several Ph.D scholars and trained solar energy experts. Two excellent solar passive houses have been constructed at the MBM Engineering College Jodhpur by Prof M. L. Mathur of MBM Engg. College, Jodhpur in collaboration with IIT Delhi. Some work on solar air heaters of Prof. Karwa of MBM Engg. College, Jodhpur is also worth mentioning. Prof. Dashora also attempted to work on material aspects of solar cooker at Jaipur. BIS published ISI standards of hot box type solar cookers and flat plate collectors. But as far as practical solar devices for agriculture is concerned, CAZRI's contribution was always significantly more.

Programmes on solar energy were initiated during that period at IIT Bombay also with Prof. S.P. Sukhatme as leader. Their work on solar air-conditioning in late 70s and early eighties is worth mentioning. Later Prof. Nayak, Dr. Kedare and team contributed quite a lot to developing standards for concentrators. A solar dryer was also developed there. Dr Rangan Banerjee organized one of the better conferences in the year 2000 in which practically all stalwarts delivered lectures. Earlier Prof. R. L. Shawney and now Prof. S. P. Singh at Devi Ahilya University Indore has taken up towards developing standards of rest of the components of solar water heaters. SPERI at Anand, Gujrat also facilitated in testing of solar devices and were earlier active in the field. IIT Madras has also been attempting on solar air conditioning and distillation. TERI (The Energy Research Institute) in collaboration with Australians installed a hybrid solar air conditioning plant. This organization has been doing a lot of surveys related to energy and potential assessment of solar energy and augmenting dissemination of solar lamps. Earlier in seventies and eighties Tata Energy Documentation Centre at Bombay provided a lot of help to researchers in collecting research papers and documenting it. Similar efforts were made by GEDA also. Special mention need to be made of Thin Film lab of IIT Delhi with Prof. K. L. Chopra as the leader where pioneering work was carried out on thin film solar cells and selective coating and later on PV systems by Prof. V Dutta. In fact the thin film of CdS grown all over the world with solution growth method and used in large scale PV modules based on CdS- CuInGaSe2 devices was developed by Prof. Chopra's group. Dr. V. G. Bhide at NPL also did some pioneering work on CdS-CuxS solar cells, selective coatings, materials and collectors for solar energy etc. During the International Solar Energy Conference in 1978, NPL demonstrated own fabricated very good cylindrical concentrators for steam generations in the exhibition. But somehow that did not take off. We also exhibited there six different solar devices developed at CAZRI and probably were one of the better centre of attraction. NPL was given a major project on polycrystalline Si solar cells where work had been carried out earlier by Dr. G. C. Jain and later by Dr. B. K. Das and Dr. S. N. Singh. On the other hand some pioneering work was carried out on CdS -CuxS solar cells and later on CdTe based devices at Jadavpur University by Prof. S Deb and on CuInGaSe2 devices by Prof. A. K. Pal, Prof. Subhadra Chowdhry and the team, Dr Ghosh, Basu and team at IIT Kharagpur also were active on GaAs based devices. PV lighting systems and PV pumps were installed by CEL in eighties. IIT Delhi installed a tracking system for PV array at FRL, Leh to enhance the PV output. One ITBP hospital was solar heated by using modified Trombe's concept by IIT Delhi. Several solar green houses were installed in cold areas. The potential of solar PV was evaluated for pumping water in India by Ramana Rao and Pande way back in 1982. CAZRI provided expert advice to army in 1990 and demonstrated PV pump based drip system in 1996 for growing pomegranate orchard in arid region. A report on feasibility of 35 MW solar thermal power plant at Mathania was also prepared by CAZRI (Gupta and Nahar 1988).

Dr H. Saha of Jadavpur University developed special inverter and solar lantern which was adopted by Ministry for providing to users during early nineties on subsidized rates. Whereas on amorphous silicon solar cells it was Prof K. L. Chopra, Prof L. K. Malhotra, Prof. D. K. Pandya at IIT Delhi and Prof. A. K. Barua, Prof. Swati Ray group at Indian Institute of Cultivation of Sciences, Calcutta who led to develop technology for scaling up. One experimental plant for commercially producing amorphous silicon devices based on Glass Tech technology of USA was installed and looked after by BHEL. Prof. S. R. Dhariwal of JNVU, Jodhpur did good work on modeling of amorphous Si based solar cells. SSPL, New Delhi also did some work on porous silicon. Dr. Patil reported to develop technology for solar cells based on organic semiconductors at IISC Bangalore, Dr. G. D. Sharma of JNVU and Dr. Roy of DLJ worked on TiO2 based devices and provided some good results. There are immense contributions from several scientists and researchers from all over the country and it is not possible to list them all, but with several projects sponsored by ministry, departments, state governments, nodal agencies mushrooming growth took place on small to medium sized projects on both solar thermal and photovoltaic devices and it is very difficult to categorically enlist those results which may be considered of noticeably better quality as far as R&D output is concerned.

Interestingly in this context, the vision of Mr. C. S. Christian, an Australian, is exemplary, who way back emphasized specifically the need of harnessing wind power and solar energy in arid region to supplement the energy needs in his famous Christian Report, on the basis of which CAZRI was established in 1952. However, it took some time and after the recommendation of fourth quinquennial review team the Division of Wind Power and Solar Energy Utilization was created in CAZRI in 1972 as mentioned earlier. Dr A. Krishnan was first Head of Division and Dr. H. P. Garg joined as Physicist. Soon the group got strength with joining of K. P. Thanvi (1975), P. C. Pande (1976), N. M. Nahar (1977), P. B. L. Chaurasia (1979), K. S. Malhotra (1980-1983), J. P. Gupta (1980) and later A. K. Singh (1990), P. Santra (2010) and off late Dilip Jain (2011) and S. Poonia (2015) and of course with inputs of Harpal Singh (1996), Dinesh Mishra (1986) and P. K. Malaviya (1989) from agricultural engineering considerations and BV Ramamna Rao, Y. S. Ramakrishna from leadership point of view. Dr H. S. Mann was one Director at CAZRI who whole heartedly supported solar energy work and facilitated the research. Dr K.A. Shankarnaryanan, Dr. J. Venketshwarlu, Dr A.S. Faroda, Dr Pratap Narain, Dr K.P.R. Vittal also need special mention who provided visionary approach as Directors for harnessing solar energy. Dr. M. M. Roy as Director and Dr. U. Burman I/C ITMU put their efforts to disseminate the developed solar technologies.

Earlier research work in India was confined to water heating, cooking, drying, distillation, space heating, cooling and power generation through solar thermal route. During late seventies and early eighties several institutes like CEERI Pilani (Prof. K. S. Rao), CAZRI, PAU, CS&MCRI etc. were engaged in the development of solar thermal pump as there was an award of Rs. 10000/- for the same. Few French solar thermal pumps were installed in UNDP sponsored project during early eighties at various sites of African countries but at a prohibitively high cost and low efficiency. As mentioned above that with emphasis on renewable energy sources due to problems created by oil crisis, different institutes and universities got engaged with solar energy research works. Both basic and applied researches have been carried out at different organizations such as Central Arid Zone Research Institute (CAZRI), Defence Laboratory (DL), MBM Engineering College and Jai Narain Vyas University (JNVU) at Jodhpur. Field Research Laboratory at Leh (renamed now as Defence Institute of High Altitude Research) has also been active in harnessing solar energy for cold desert (Rao 1984). Research work was carried out at IITs, NPL, IISc Bangalore, IICS Calcutta, Jadavpur University Calcutta, Kamraj University Madurai, TNAU, Coimbatore, PAU Ludhiana, CTAE, Udaipur, TERI etc. Some worth mentioning results of the research outcome at CAZRI are enlisted, which had more implications as far as applications of solar energy for agriculture is concerned.

Basic work at CAZRI: Salient points

- Collector-cum-storage solar water heater reduces the cost to half of the conventional solar water heater (Garg 1975).
- Collector reflector geometry was optimized with width to length ratio as three to four to develop stationary solar devices (Pande et al. 1978; Pande and Thanvi 1988 a).
- In solar stills basic design and operational parameters were optimized. Roof angle of 10-15^0 in double sloped solar still is preferable (Garg and Mann 1976).
- Mirror boosters can enhance 20% distillate output in single sloped solar still (Thanvi and Pande 1980).
- Thermal losses in solar still were quantified indicating need of insulation at the rear side (Thanvi and Pande1988b).
- Reflectors of extended length were found to enhance PV output by 17-20% with additional cost of 2-3 % on reflector (Pande and Dave 1999, 2007).
- Reduction in the current output of PV panels due to dust on panel ranged from 15-30 % in open field (Pande 1992a) if panels were not cleaned, indicated need to have a regular cleaning schedule of PV modules.
- The gap between absorber and cover glazing should be 40-50 mm to reduce convective thermal losses with minimal shade effect in solar (Nahar and Garg 1981)
- Natural circulation type water heater with flat plate collector using GI pipes header and riser and aluminum sheet as absorber save 30% cost while giving the comparable performance compared to commercially available copper pipes and copper sheet flat plate collectors (Nahar and Gupta 1985).
- Al_2O_3-Co selective coatings on aluminum substrate and black nickel coatings on stainless steel were found effective (Nahar et al. 1989).
- Studies on passive method by incorporating embedded pipes in cement concrete slabs indicated suitability of the system in providing hot water at moderate temperatures above 40°C. (Chaurasia 2000).
- In solar dryers dried pearl millet stalk was found to be a convenient insulation material (Thanvi and Pande 1987, 1988).
- The effect of dust on the transmittance of glass and deterioration in specular reflectance of mirror at Mathania ranged from 24 to 50% (Gupta et al. 1994).
- CdS-CdTe thin film devices were prepared by electrophoresis process using polar organic solvents as dispersion media and could be made robust through laser induced recrytallization (Pande 1994; Pande et al. 1996, 1997)
- Orientation schedule of PV array in solar pump was standardized for getting higher output.

Solar Devices in Past

The basic results led to the development of several solar energy gadgets. Some of solar devices relevant for rural and agricultural applications are mentioned to have a feel of spectrum.

Solar dryers

- At CAZRI, solar cabinet dryer (Lawand design) was developed (Garg and Krishnan 1974) and subsequently improved dryer with ATRM was evolved (Pande 1980). In eighties, these solar cabinet dryers were used by local entrepreneurs for dehydrating catechu/sugar coated aniseeds and tobacco powder.
- PAU Ludhiana developed solar dryer for chillies and installed at village Ladowal in 1977.
- Inclined solar dryer was developed at CAZRI (Pande and Thanvi 1982) followed by installation of improved and scaled up versions up to1000kg capacity in village Keru.
- Solar cabinet dryer with aspirator at CIAE Bhopal (Singh, H. and Alam, A 1983)
- Solar air heaters and forced circulation type solar dryer were developed at CAZRI (Pande et al.1979, Garg et al. 1980; Pande 1980a,b) followed by mixed mode dryer.
- Multi rack solar dryer and packed bed air heaters were developed at PAU Ludhiana and CAZRI Jodhpur.
- Solar dryers of different designs at IIT Delhi, IIT Bombay, TNAU, Coimbatore and many institutes
- CTAE Udaipur developed solar tunnel dryers for large scale drying (Rathore and Mathur 2002).
- Simple solar dryers were extensively used in Pune with standardized processes for drying grapes and aonla.
- Solar dryer -cum -water heater was developed at CAZRI (Pande and Thanvi, 1991). This dual purpose device is capable to continue the dehydration process even at night. It efficiently dehydrated different grasses for making hay (Mali et al. 1999)
- Watermelon candies were prepared in three in one integrated solar device by using the device first in cooking mode and then drying with built in thermal storage facility (Pande 2009)
- Mixed mode solar PV dryer was developed (Pande et al. 2010) and its scaled up version was installed at KVK Pali
- Solar dryer with PCM storage material has been experimented successfully (Jain, D. and Tewari, P. 2015)

- Photovoltaic thermal (PV/T) hybrid solar dryer with PCM storage material has been experimental successfully (Poonia et al. 2018 a,b).

Solar dryers are available in different parts of country with some changes in design to suit the crop and location.

Solar desalination

- Low glass roof type solar stills (Garg and Mann 1976).
- Large size solar still by CSM & CRI, Bhavnagar, installed at village Avania (Gujarat) and Bhaleri (Churu, Rajasthan)
- Improved Solar stills with better sealing and incorporation of a system to prevent algae and scale formations (Gupta et al. 1990)
- Step basin tilted type solar stills (Thanvi 1985, Thanvi and Pande 1989, 1990). Railways, electric grid stations, army units and schools have adopted such multi basin tilted type solar stills
- Defence Lab, Jodhpur developed PV operated electro-dialysis plants and installed at Barmer to provide potable water to villagers.
- A solar water pyramid innovated by Martijn Nitzsche of Netherlands, has been installed at Roopaji Raja Beri in Barmer district by Jal Bhagirathi Foundation, Jodhpur. (Sebastian 2009).
- Solar stills were developed with different construction materials (Nahar et al. 2013 and Singh et al. 2019)

Solar cookers

- Simple hot box type solar cookers with single reflector are commercially available in market.
- BIS published Indian Standards for solar cookers. State nodal agencies promoted such cookers in nineties.
- A simple solar hot box cooker with masonry structure was in use during early nineties at SOS school, Leh.
- Some snow melting ovens were developed by DLJ.
- At CAZRI solar oven (improved version of Telkes cooker) was found better among five solar cookers (Garg et al. 1978a).
- Double mirror box type solar cooker developed by (Gupta and Purohit 1986) requires tracking after only three hours.
- Stationary cooker developed at CAZRI (Pande and Thanvi 1987, 1988 a) with optimized width to length ratio can be used without sun tracking and utilizes maximum solar energy due to its especial geometry.
- Non-tracking solar cookers, both for domestic and community purposes, were developed (Nahar et al.1993) at CAZRI

- Solar cooker with tracker was developed at CAZRI (H. L. Kushwaha)
- Solar cooker for boiling animal feed found more acceptability (Nahar et al.1994a). Such cookers were also experimented at CTAE, Udaipur.
- Community cooking through concentrators has been successfully demonstrated at Brhmakumari Ashram, Mount Abu. Such Scheffler cookers (named after the German inventor) have also been installed at Tirupati and Sirdi temples.
- A solar cafeteria has been installed in Bhuj of Kachchh region in Gujarat and reported to be working satisfactorily (Sharan and Mania 2005).
- Concentrator based cookers are being promoted by MNRE but need of continual attention is one of the impediments.

Novel Integrated Solar Devices developed at CAZRI

- Solar water heater-cum -steam cooker (Garg et al. 1978)
- Multipurpose solar device (Pande et al. 1981)
- Solar cooker-cum-dryer (Pande and Thanvi 1988a)
- Step basin type solar water heater-cum-solar still (Thanvi and Pande 1988b)
- Solar water heater-cum-dryer (Pande and Thanvi 1991)
- Solar integrated device for water heating, cooking and drying (Pande 2003, 2006 and 2009a).

Environmentally controlled enclosures for growing plants

- Earth tube cooled green houses at Kachchh, Bhuj (Sharan and Jadhav 2002)
- Solar green house (IIT Delhi, CTAE Udaipur, CAZRI, Jodhpur, PAU Ludhiana and many Institutes)
- PV clad enclosure with earth embedded pipes for environment control, developed at CAZRI (Pande et al. 2013)

Solar devices for income generation at CAZRI

- Solar candle machine (Chaurasia et al.1983)
- Solar polish making machine (Pande 1999)
- Solar still for rose water (Thanvi and Pande 1989d)

Solar Photovoltaic Applications for Agriculture at CAZRI

- Solar PV pump- based drip system for growing orchards (Pande et al. 2003)
- PV Generator for multiple applications (Pande et al. 2008)
- Solar PV sprayer CAZRI, Jodhpur (Pande 1990)
- Solar PV duster, CAZRI, Jodhpur (Pande 1998)

- Solar PV winnower (Pande 2003a)
- PV winnower -cum –dryer (Pande et al. 2008; Pande 2009b)
- PV mobile unit (Pande 2009c)

Present and Future

During first decade of this century, with prices of solar cells coming down and projections indicating favourable situations, more emphasis has been laid on application of PV and large scale applications. National Solar Mission was launched and initially a target was set for 22,000 MW solar power installations by 2022. This is now enhanced to 100,000 MW. Several solar power plants have been commissioned and several are in pipeline. Roof top installations are given preference. A small PV roof system was installed at ATIC in CAZRI. All attentions are now on smart grids, inter phasing and stand alone to grid connected larger systems. Debate on PV vs Solar thermal keeps on cropping up. Solar PV pumps are installed with huge target of 100, 000 or more.

Development of PV structures for providing environmental control to raise nursery and use it for growing high value crops in off season has a good scope. Considering the need of optimization of PV pump based drip systems for different capacities depending on water harvesting structures and command area at CAZRI. Green buildings with hybridization of renewable energy sources, use of storage materials and materials studies for providing low cost, robust, durable and practically adoptable devices need to be given more emphasis. Need is to evolve a better low cost storage batteries, alternatives to storage system, development of low cost efficient quantum dot solar cells, measures to do away dust problem and integrated combo devices for more agricultural and practical applications.

The dissemination and diffusion has been a concern. Not withstanding the economic impediments, one of the major problems is non availability of these solar devices in market. Efforts were made at CAZRI to develop a module for the transfer of technology so that entrepreneurs could take up and produce commercially the devices. Technology of five solar devices has already been provided to local entrepreneurs. National Solar Mission envisages installation of several solar power plants in phased manner towards achieving set goals of 100,000 MW by 2022 and simultaneously providing the energy needs through solar devices. Although large number of companies has diverted their resources towards the big solar power plants, creating R & D facilities also, the scope of disseminating small and medium size multi purpose integrated solar thermal and PV device is enormous. Incentives for establishment of cottage industries in villages by local skilled youth may facilitate not only the development of these devices but would promote self employment.

Finally lead has been taken by CAZRI in the development of solar farms with possible use of inter space of PV arrays for raising economically viable crops. This will have long implications for both farmers and solar power developers. Concerted efforts are needed for the development of a comprehensive solar package.

References

Chaurasia, P.B.L. 2000. Solar water heaters based on concrete collectors, International Journal of Energy, 25: pp.703-716.

Chaurasia, P.B.L., Gupta, J.P. and Ramana Rao, B.V. 1983. Comparative study of performance of two models of solar device for melting wax during winter season. Energy Conversion and Management 23(2): 73-75.

Garg H.P. 1975. Year round performance studies on a built in storage type solar water heater at Jodhpur. Solar Energy 17(3): 167-172.

Garg, H.P. and Krishnan, A. 1974. Solar drying of agricultural products Part I: Drying of chillies in a solar cabinet dryer. Annals of Arid Zone 13(4): 285-292.

Garg, H.P. and Mann, H.S. 1976. Effect of climatic operational and design parameters on the year round performance of single slope and double slope solar still under Indian arid zone conditions. Solar Energy 18 (2): 159-174.

Garg, H.P., Mann, H.S. and Thanvi, K.P. 1978a. Performance evaluation of five solar cookers. Proceedings of International Solar Energy Conference, New Delhi, pp. 1491-1494.

Garg, H.P., Pande, P.C. and Thanvi, K.P. 1980a. Design and development of a fruit and vegetable agricultural dryer. Proceedings and selected papers: International Symposium on Arid Zone Research and Development (Ed. H.S. Mann), Scientific Publishers, Jodhpur, pp. 405-410.

Gupta, J.P. and Nahar, N.M. 1988. Feasibility of 30 MW Solar Thermal Power Station at Jodhpur. Urja 23: 468-470.

Gupta, J.P. and Purohit, M.M. 1986. Role of renewable energy sources for mitigation of cooking fuel problem. Trans. Indian Society Desert Technology University Centre Desert Studies 11(1): 7-17.

Gupta, J.P., Sharma, P. and Purohit, M.M. 1990. Improved method for pre-treatment of brackish water for solar distillation to prevent algae growth and scale formation. Renewable Energy and Environment. Proceedings of National Solar Energy Convention, Dec. 1-3, 1989 (Eds. A.N. Mathur and N.S. Rathore), Himanshu Publication, Udaipur, pp. 33-36.

Gupta, J.P., Sharma, P., Nahar, N.M., Kackar, N.L. and Purohit, M.M. 1994. Design of suitable green house for arid areas. Proceedings National Solar Energy Convention, Gujarat Energy Development Agency, Vadodara, pp. 123-126.

Jacobson, A. 2000. Solar Energy Managements for Ladakh, India. Renewable Energy Technology for the New Millennium. Proceedings of 24th National Renewable Energy Convention 2000. Allied Publishers Ltd. pp 5-10.

Jain, D. and Tewari, P. 2015. Performance of indirect through pass natural convective solar crop dryer with phase change thermal energy storage. Renewable energy 80: 244-250.

Mali, P.C., Pancholi, R., Mathur, B.K. and Pande, P.C, 1999.Effect of drying methods on the chemical composition of hay made from two desert grasses. Tropical Grasslands 33 (1999): 51-54.

Mani, A. and Rangarajan, S. 1982. Solar Radiation over India. Allied publisher Pvt. Ltd., New Delhi, 646 pp.

Nahar N.M., Singh A.K., Sharma P. and Chaudhary G.R., 2013. Design development and performance of solar desalination device for rural arid areas, In: Proceedings of International Conference on Renewable Energy(ICORE) 2013 (Eds. M. Kumaravel, S.M. Ali, S.K. Samdarshi, Ranjan Jha, Jagat S. Jawa), Excel India Publishers and SESI, New Delhi, pp. 50-52.

Nahar, N.M. and Gupta, J.P. 1985. Performance and testing of an.improved natural circculation type solar water heater. Energy Conversion and Management 27(I): 29-32.

Nahar, N.M. and Gupta, J.P.1989. Studies on gap spacing between absorber and cover glazing in flat-plate solar collectors. International Journal of Energy Research 13: 727-732.

Nahar, N.M., Gupta, J.P. and Sharma, P. 1993. Performance and testing of an improved community size solar cooker. Energy Conversion and Management 34: 327-333.

Nahar, N.M., Gupta, J.P. and Sharma, P. 1994a. Design development and testing of large size solar cooker for animal feed. Applied Energy 48: 295-304.

Pande P.C., Singh A.K., Ansari S. and Dave B.K 2008. PV generator for multiple applications in arid farming. In Diversification of Arid Farming Systems (Eds Pratap Narain, M.P.Singh, Amal Kar, S. Kathju and Prvaeen Kumar) CAZRI & Scientific Publishers (India) Jodhpur, pp. 273-278.

Pande P.C., Singh A.K., Ansari S., Vyas, S.K. and Dave B.K. 2003. Design development and testing of a solar PV pump based drip system for orchards, Renewable Energy 28:385-396.

Pande P.C., Singh A.K., Purohit, M.M. and Dave B.K. 2008. A mixed mode solar PV dryer. Proceedings World Renewable Energy Congress, Glasgow, U.K. pp. 1746-1751.

Pande, P.C. 1980. Drying Fruits and Vegetables using solar technology. Indian and Eastern Engineer 122(7): 303-305.

Pande, P.C. 1980a. Performance studies on solar air heater with different design parameters. Proceedings National Solar Energy Convention, Annamalainagar, Allied Publishers, pp. 90-95.

Pande, P.C. 1980b. Performance studies on an improvised solar cabinet dryer. Proceedings National Solar Energy Convention, Annamalainagar, Allied Publishers, pp. 1-5.

Pande, P.C. 1990 b Development of PV systems for arid region. In Energy and the Environment into the 1990s. Proceedings 1st World Renewable Energy Congress (Ed. A.A.M. Sayigh), Pergamon Press, Oxford, I: 314-318.

Pande, P.C. 2003c. Design and Development of Solar Technologies for Arid Region. In : Impact of Human Activities on Thar Desert Environmen. Eds. Pratap Narain, S. Kathju, Amal Kar, M.P. Singh, Praveen Kumar. Arid Zone Association of India and Scientific Publishers, Jodhpur. pp. 554-565.

Pande, P.C. 2009 c. A PV Mobile Unit for Multipurpose Rural Applications. Proceedings International 18th Photovoltaics Solar Engineering and Science Congress, Calcutta. Eds. Swati Ray and P. Chatterjee. McMilllan Press, India. pp. 398- 403.

Pande, P.C. and Dave, B.K. 1999, Effect of boosters on the performance of PV panels. In: Renewable Energies and Energy Efficiency for Sustainable Development. Proceedings 23rd National Renewable Energy Convention, Dec. 20- 22, 1999, Devi Ahilya University, Indore (Eds. R.L.Shawney, D.Buddhi and R.P.Gautam) pp.120-121.

Pande, P.C. and Dave, B.K. 2007 Economical Production of Electricity from PV Mod ules for Application in Post-Harvest Operations. In: Advances in Energy Research. Porceedings 1st International Conference on Advances in Energy Research, IIT Bombay. Macmillan India Ltd. pp 296-300.

Pande, P.C. and Gupta, J.P. 1997. Management of Energy Resources. In: Desertification control in the arid ecosystem of India for sustainable development. (Eds. Surendra Singh and Amal Kar), Agro Botanical Publisher (India), Bikaner, pp. 359-370.

Pande, P.C. and Thanvi, K.P. 1982. Solar dryer for maximum energy capture. Proceedings NSEC, Allied Publishers, New Delhi, pp. 400.9-400.12.
Pande, P.C. and Thanvi, K.P. 1987. Design and Development of a solar cooker for maximum energy capture in stationary mode. Energy Conversion and Management 27(1): 117-120.
Pande, P.C. and Thanvi, K.P. 1988a. Design and development of solar cooker-cum-dryer. International Journal of Energy 12: 539-545.
Pande, P.C. and Thanvi, K.P. 1988b. Practical evaluation of integrated solar energy gadgets. Energy option for 90's Proceedings NSEC. lIT New Delhi (Eds. N.K. Bansal, V.V.N. Kishore and Anil Mishra), Tata McGraw Hill Publishing Co. Ltd., New Delhi, pp. 178-183.
Pande, P.C. and Thanvi, K.P. 1991. Design and development of solar dryer cum water heater. Energy Conversion and Management 31 (5): 419-424.
Pande, P.C. I 992a. Effect of dust on the performance of PV panel. Proceedings 6th International Conference on Photovoltaic Science & Engineering (Eds.B.K.Das and S.N. Singh), Oxford and IBH Publishing Co., pp.539-542.
Pande, P.C., 1998. A novel solar device for dusting insecticide powder. Proceedings of National Solar Energy Convention, Univ. of Roorkee, Roorkee. pp 117-122.
Pande, P.C., 2006. Design and development of PV winnower -cum -dryer. Proeedings. International Congress on Renewable Energy (ICORE 2006). Eds. E.V.R. Sastry and D. N. Reddy. Allied Publishers Pvt. Ltd., pp.265-269.
Pande, P.C., Bocking, S., Miles, R.W., Carter, M. J., Latimer, J.D. and Hill, R. 1996. Recrystallization of electrophoretically deposited CdTe films. Journal of Crystal Growth 159: 930-934.
Pande, P.C., Bocking, S.,Miles, R.W., Carter, M.J. and Hill, R. 1997. Laser Induced Recrystallisation of Electrophoretically Deposited CdTe and CuInSe2 Films. Solid State Phenomena. 55: 146-148.
Pande, P.C., Garg, H.P. 1978. Review of thermal storage materials from the view point of solar energy application. Proceedings National Solar Energy Convention, Bhavnagar, Allied Publishers, New Deihl; pp.191-202.
Pande, P.C., Garg, H.P. and Thanvi, K.P. 1979. Performance studies on solar air heaters for the development of solar fruit and vegetables dryer. Proceedings National Solar Energy Convention, Bombay, pp. 23-28.
Pande, P.C., Singh, A.K., Dave, B.K. and Purohit, M.M. 2010. A preheated Solar PV dryer for economic growth of farmers and entrepreneurs. SESI Journal, 20 (1&2):73-79.
Pande, P.C., Singh, A.K., Santra, P., Vyas, S.K., Purohit, M.M., Dave, B.K., 2013. Studies on PV clad structure for controlled environment. In: Proceedings of International Conference on Renewable Energy (ICORE) 2013 (Eds. M. Kumaravel, S.M. Ali, S.K. Samdarshi, Ranjan Jha, Jagat S. Jawa), Excel India Publishers and SESI, New Delhi, pp. 294-297.
Pande,P.C. 2009.a Watermelon Processing in Three in one Integrated device Proceedings of International Solar Food Conference, Indore, pp. 26.
Pande,P.C. 2009b. Performance of PV winnower cum dryer for processing of agricultural products. Proc. International Solar Food Conference, Indore, pp. 27
Pande. P.C. 1994. A study on electrophoretically deposited CdS-CdTe devices. In Role of Renewable Energy in Energy Policy (Eds. K.M. Dholakia, M.M. Pandey, N. Gandhi and D. Vaja), Gujarat Energy Development Agency, Vadodara,pp.140-144.
Pande. P.C. 2003a. A Solar PV winnower. Proceedings 26th National Renewable Energy Convention and International Conference in New Millennium, Coimbatore. 2003. pp. 22-26.
Poonia, S; Singh, A.K. and Jain, D. 2018. Design, development and performance evaluation of photovoltaic/thermal (PV/T) hybrid solar dryer for ber (Zizyphus mauritiana) fruit. Cogent Engineering 5(1): 1-18. DOI:10.1080/2331916.2018.150784.

Poonia, S; Singh, A.K. and Jain, D. 2018. Mathematical modelling and techno-economic evaluation of hybrid photovoltaic-thermal forced convection solar drying of Indian Jujube. Journal of Agricultural Engineering 55(4): 74-88.

Rao S.K. 1984. An oasis in the cold Desert- Field Research Laboratory, Leh. Def. Sc. Journal. 34(2): 213-220.

Rathore, N. S., and Mathur, A.N. 2002 Design and development of solar tunnel dryer for industrial use. Proceedings of All India Seminar on Advances in Solar Technologies, Jodhpur. Jan. 29-30, 2002. Editor Rajendra Karwa. pp. 189-196.

Sebastian, S. 2009. A water wonder in the middle of a desert. http://www.thehindu.com. /2009/03/06/stories/2009030656090800.htm

Sharan, G. and Jadhav,R. 2002. Greenhouse cooled by earth tube heat exchanger at Kothara (Kutch). Alumunus 34(1): 33.

Singh, A.K. Poonia, S., Jain, D. and Mishra, D. (2019). Performance evaluation and economic analysis of solar desalination device made of building materials for hot arid climate of India. Desalination and Water Treatment 141(2):36-41) DOI: 10.5004/dwt.2019.23480

Singh, H and Alam, A. 1983. Development of Solar-cum wind spirator for drying and ventilation. Journal of Agriculture Enginnering 14(4):63-68.

Thanvi, K.P. 1985. Performance studies on family size solar stills in arid zone of India. Proceedings U.S. India symposium on solar energy research and applications, Univ. of Roorkee, pp. 291-294.

Thanvi, K.P. and Pande, P.C. 1980. Effect of boosters on the performance of solar stills. Proceedings National Solar Energy Convention, Allied Publishers, pp. 116-122.

Thanvi, K.P. and Pande, P.C. 1987. Development of low cost solar agriculture dryer for arid region of India. Energy in Agriculture 6: 35-40.

Thanvi, K.P. and Pande, P.C. 1988a. Designing low cost solar dryer with alternative materials. Energy options for 90's. Proceedings NSEC. IIT, New Delhi (Eds. N.K. Bansal, V.V.N. Kishore and Anil Mishra), Tata McGraw Hill Publishing Co. Ltd., New Delhi, pp. 70-74.

Thanvi, K.P. and Pande, P.C. 1988b. A step basin type solar water heater cum still. International Journal of Energy Research 12: 363-368.

Thanvi, K.P. and Pande, P.c. 1989b. A low cost tilted type solar dryer. In Renewable Energy for Rural Development, Tata McGraw Hill Publishing Co. Ltd., New Delhi, pp. 405-409.

Thanvi, K.P. and Pande, P.C. 1989c. Design and development of solar still for electric grid station. Research and Industry 34: 56-59.

Thanvi, K.P. and Pande, P.C. 1989d. Development of solar still for rose water production. In Renewable Energy for Rural Development, Tata McGraw Hill Publishing Co. Ltd., New Delhi, pp.380-384.

Thanvi, K.P. and Pande, P.C. 1990. Development of inclined solar dryer with alternate materials. Renewable Energy and Environment (Proceedings of NSEC), Dec. 1-3, 1989, Udaipur (Eds. A.N. Mathur and N.S. Rathore), Himanshu Publication, Udaipur, pp.41-45.

3

Solar Passive Techniques for Cooling in Arid Region

Dilip Jain

ICAR- Central Arid Zone Research Institute, Jodhpur, Rajasthan, India

Introduction

In Indian Thar desert regions, excessive heat is the major problem that causes human and animal thermal discomfort. Space cooling is, therefore, the most desirable factor for the inhabitants. Various examples of dwellings responsive to climatic constraints are found in vernacular architecture throughout the world. Compact cellular layout with minimum external surface exposure to the sun, whitewashed surfaces to reduce absorptivity, blind external facades, courtyards, vegetation to provide humidity and shade, and heavy buildings constructed from high thermal capacity materials are common passive features in most of the arid regions. Wind towers for cooling ventilation are well known in Iranian and Middle East architecture, which along with cooling of the air by earth and water evaporation keep the building comfortable in hot periods. Building underground to take advantage of the large thermal storage capacity of the earth is also used in Tunisia and central Turkey.

Factor affecting the heat accumulation

The main sources of heat accumulation as environmental parameters are solar heat gain, internal thermal mass, air leakage and temperature difference (Fig.3.1).

The outer surface of building, its absorptivity and orientation may be taken as the structural parameters. Thus, the passive cooling involves checking these effects and designing the structure in such a way so that heat accumulations could be eliminated.

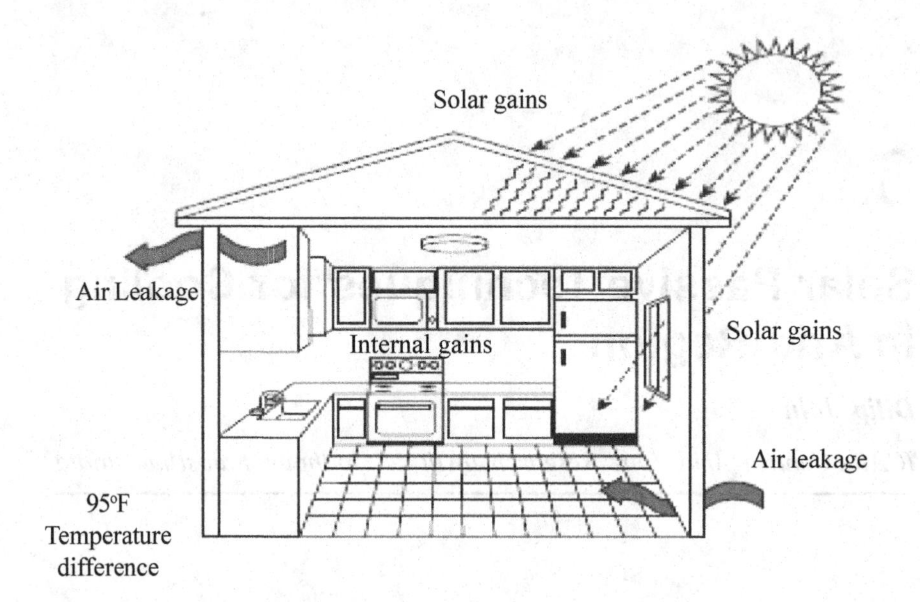

Fig. 3.1: Factors affecting heat accumulation in a house

What is passive cooling?

Passive cooling is defined as the removal of heat from the building environment by applying the natural processes of elimination of heat to the ambient atmosphere by convection, radiation and evaporation or to the adjacent earth by conduction and convection. In arid region, summer and winter are extreme in both the sides. In such situation techniques of passive cooling are much appropriate to perform job of comfort. In summer, it can protect from excess solar radiation and on the other hand the same technique can conserve the heat of the house and protect from the nocturnal cooling in the winters.

Understanding the passive cooling

Roof only covers the 36.7% of total solar radiation falling on the single storied building having all four sides exposed to sun in the summer. Many theoretical investigations also presented the thermal models of various passive cooling methods. Thermal modelling of passive cooling is important to study the design and operating parameters. Simulation models are valuable tools for prediction of performance of the passive cooling.

The periodic analysis presented for hourly variation of passive cooling of various roof treatments such as bare roof, insulation beneath the roof, evaporative cooling above the roof and roof pond with movable insulation system are cost effective to understand the whole passive cooling systems. A multifunctional solar systems are used including indoor ventilation, double walls and triple roof in order to remove heat. The external walls were clad with unglazed brick to allow evaporative cooling. Analysis showed that the different solar passive techniques *viz.* radiant cooling, evaporative cooling, white painted roof, insulated roof and roof pond with movable thermal insulation were effective in hot and dry climate. The climate chart (Fig.3.2) shows the design strategies for difference passive cooling.

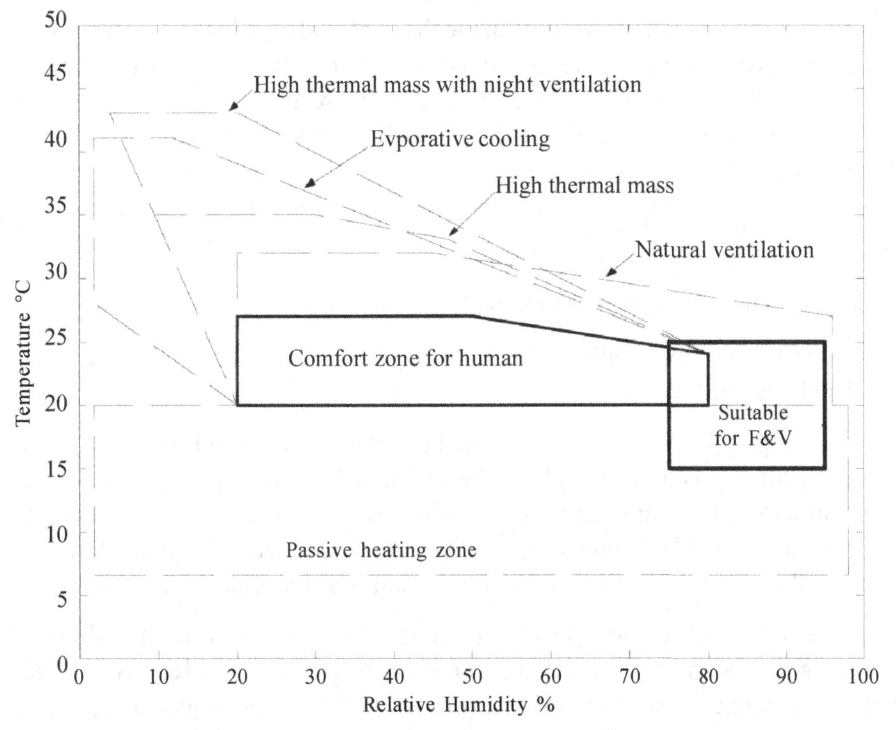

Fig. 3.2: Climate chart with design strategies for different passive cooling techniques

To prevent heat from entering into the building or to remove once it has entered is the underlying principle for accomplishing cooling in passive cooling concepts. This depends on two conditions: the availability of a heat sink which is at a lower temperature than indoor air, and the promotion of heat transfer towards the sink. Environmental heat sinks are:

- Outdoor air (heat transfer mainly by convection through openings)
- Water (heat transfer by evaporation inside and / or outside the building envelope)
- The (night) sky (heat transfer by long wave radiation through the roof and/ or other surface adjacent to a building
- Ground (heat transfer by conduction through the building envelope)

Passive cooling techniques can reduce the peak cooling load in buildings, thus, reducing the size of the air conditioning equipment and the period for which it is generally required.

Various passive cooling techniques

Thermal lag effect (thick wall)

The thickness of wall helps in shift in the thermal peak inside the building. The thick walls were being constructed since ancient as a passive cooling method. Thermal lag describes a body's thermal mass with respect to time. A body with high thermal mass (high heat capacity and low conductivity) will have a large thermal lag.

$$thermal\,lag(s) = \sqrt{\frac{1}{2*\alpha*\Omega}}*L$$

where, α = Thermal diffusivity (m^2/s^{-1})

Ω = External angular frequency (s^{-1})

L = Thickness (m)

The slow night-time cooling of a home after its external brick wall has been heated by the sun is one example of thermal lag. Thermal lag is the reason the high temperatures in summer continue to increase after the summer solstice (in this case, it is termed seasonal lag), and it is the reason a day's high temperature peaks in the afternoon instead of when the sun is at its peak (12 noon).

Ventilation: In recent years, several investigations were performed and showed that there can be multiple solutions to the excessive heat problem. A popular method is cooling through ventilation. Ventilation is very common and important technique of passive cooling. Ventilation can be achieved by different ways, however can be defined mainly as i) natural ventilation and ii) wind induced ventilation. The natural ventilation occurs due to temperature difference and thermal buoyancy. The thermal buoyancy-driven air volumetric flow rate (m^3s^{-1}) with two small opening ventilation in natural convection is

$$Q_a = \frac{C_d A}{\sqrt{1+a}}\sqrt{\frac{2gH(T_2 - T_1)}{T_1}}$$

where, C_d = Coefficient of diffusivity

g = Gravitational acceleration ($m\ s^{-2}$)

H= Height of vent (m)

A = Cross sectional area of vent (m^2)

a = Ratio of cross sectional area of outlet and inlet of air flow channel

T_1 and T_2= Temperature in inside and outside structure respectively

The thermal buoyancy-driven air volumetric flow rate ($m^3 s^{-1}$) with large single opening ventilation can be expressed as

$$Q_a = \frac{C_d A}{3}\sqrt{\frac{gH(T_2 - T_1)}{T_1}}$$

The wind induced ventilation depends upon the velocity of wind and area of opening in building. The empirical relation of ventilation due to wind is expressed as

$$Q_a = 0.025A.V$$

where, V is wind velocity (m s^{-1}).

Evaporative cooling: Another most common practice of passive cooling is evaporative cooling. Evaporative cooling can be effectively achieved at the higher difference in temperature of dry bulb and wet bulb. It is reduction in temperature resulting from the evaporation of a liquid, which removes latent heat from the surface from where evaporation takes place. It is an adiabatic process obtained at constant enthalpy. The evaporative cooling can be understood on Psychrometric chart (Fig. 3.3). Investigation of evaporative cooling in combination with ventilation, wind induced and roof and wall treatments has proved to be the most effective and shown promising results.

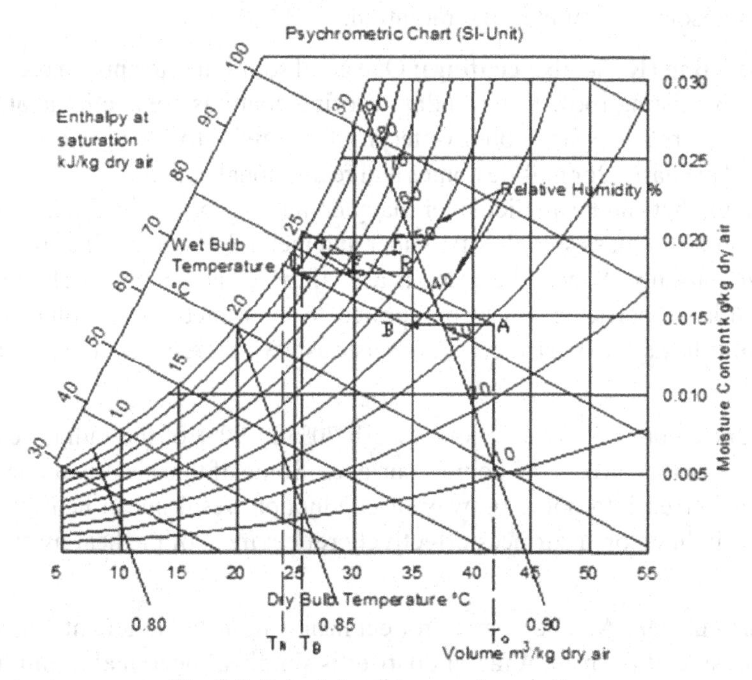

Fig. 3.3: Principle of evaporative cooling

Under an evaporative cooling, the maximum possible temperature drop is possible to the wet bulb temperature i.e. wet bulb depression. However, the evaporative cooling works normally on 80-90% efficiency. The direct saturation efficiency (ε) can be determined as follow:

$$\varepsilon = \frac{T_{e,db} - T_{l,wb}}{T_{e,db} - T_{e,wb}}$$

where, $T_{e,db}$ = Entering air dry-bulb temperature (°C)

$T_{l,wb}$ = Leaving air dry-bulb temperature (°C) and

$T_{e,wb}$ = Entering air wet-bulb temperature (°C)

The evaporative cooling approach for passive cooling of buildings roof in hot arid climates has also become an attractive subject of investigation. However, the relative advantages of evaporative cooling in relation to many other approaches such as cavity wall, insulation, whitewash and large exposure orientations, vegetable pergola shading, roof with removable canvas, water film, soil humid grass and roof with white pots as cover also needs to be compared.

Roof cooling: About a third of the unwanted heat that builds up in home comes in through the roof. This is hard to control with traditional roofing materials. For example, unlike most light-colored surfaces, even white asphalt and fibreglass shingles absorb 70% of the solar radiation.

Roof painting (Reflective coating): One good solution is to apply a reflective coating to existing roof. Two standard roofing coatings are available at local hardware store. They have both waterproof and reflective properties and are marketed primarily for mobile homes and recreational vehicles. One coating is white latex that can be applied over many common roofing materials, such as asphalt and fibreglass shingles, tar paper and metal. Most manufacturers offer a 5-year warranty. A second coating is asphalt based and contains glass fibers and aluminum particles. It can be applied to most metal and asphalt roofs. Because it has a tacky surface, it attracts dust, which reduces its reflectivity somewhat.

Dull, dark-colored home exteriors absorb 70% to 90% of the radiant energy from the sun that strikes the home's surfaces. Some of this absorbed energy is then transferred into home by way of conduction, resulting in heat gain. In contrast, light colored surfaces effectively reflect most of the heat away from the home.

Radiant barrier: Another way to reflect heat is to install a radiant barrier on the underside of the roof. A radiant barrier is simply a sheet of aluminium foil with a paper backing. When installed correctly, a radiant barrier can reduce heat gains through the ceiling by about 25%.

Shading or blocking the heat: Two excellent methods to block heat are insulation and shading. Insulation helps keep home comfortable and saves money on mechanical cooling systems such as air conditioners and electric fans. Shading devices block the sun's rays and absorb or reflect the solar heat. The results showed that cooling ventilation using a solar chimney can reduce internal temperature of buildings. Shading devices (overhangs and verandas) to reduce summer solar radiation are effective with appropriate depths of these shading elements for various orientations.

Roof insulation: Weatherization measures such as insulating, weather stripping, and caulking help seal and protect the house against the summer heat in addition to keeping out the winter cold.

Roof pond and insulation: The inside air temperature can be reduced to within 1°C of the ambient temperature using evaporative cooling, a solar chimney or a water pond with movable insulation. Providing thermal insulation underneath the roof is less efficient than putting the thermal insulation above the roof. The performance of evaporative cooling is best, but it requires a considerable amount of water, which is scarce in rural arid areas. A shallow pond with movable thermal insulation over the roof provides better conditions inside metal structures in hot location. This technique is also quite useful in winter and at night time. The use of solar chimney (solar buoyancy) is an effective technique for reducing the temperature inside the structure. It provides also ventilation, which helps in lowering the humidity and achieving comfortable conditions inside the space.

Space cooling can also be achieved by improving the performance of roofs. This is because the roofs are the surfaces most exposed to direct solar radiation and can cause excessive heat gain in hot periods. Some efforts were made by investigators to improve roof thermal performance. The use of low emissivity material in the attic of a building reduced the underside ceiling surface temperature, which lowered the room air temperature.

The reduction of heat gain through the roofs using evaporative cooling systems was extensively investigated with open roof ponds on water spraying over the roof, moving water layer over the roof, thin water film and roofs with wetted gunny bags. The influence of evaporative cooling over the roof as compared to the bare roof case and intermittent ventilation as compared to the continuous or no-ventilation case have been assessed for controlling the indoor air temperature. It was found that the effectiveness of the evaporative cooling can be improved by conscious choice of the rate and duration, which controls the inside air temperature significantly. It was found that a combination of evaporative cooling and variable ventilation can make the internal environment of a building more comfortable. The roof pond system with stationary water is more effective in stabilizing the fluctuations of indoor temperature.

The average reduction in inside air temperature in an increasing order is found to be 6.5, 7.0, 8.5, 9.3, and 9.6°C for roof with white paint, insulated roof, solar thermal buoyancy, roof provided with shallow pond with movable thermal insulation and evaporative cooling.

Earth coupled cooling: This technique is used for passive cooling as well as heating of buildings, which is made possible by the earth acting as a massive heat sink. At depths beyond 4 to 5m, both daily and seasonal fluctuations die out and the soil temperature remains almost constant throughout the year. Thus, the underground or partially sunk buildings will provide both cooling (in summer) and heating (in winter) to the living space. A building may be coupled with the earth by burying it underground or berming.

Earth air tunnel: The use of earth as a heat sink or a source for cooling/heating air in buried pipes or underground tunnels has been a testimony to Islamic and Persian architecture. The air passing through a tunnel or a buried pipe at a depth of few meters gets cooled in summers and heated in winters (Fig. 3.4). Parameters like surface area of pipe, length and depth of the tunnel below ground, dampness of the earth, humidity of inlet air velocity, affect the exchange of heat between air and the surrounding soil.

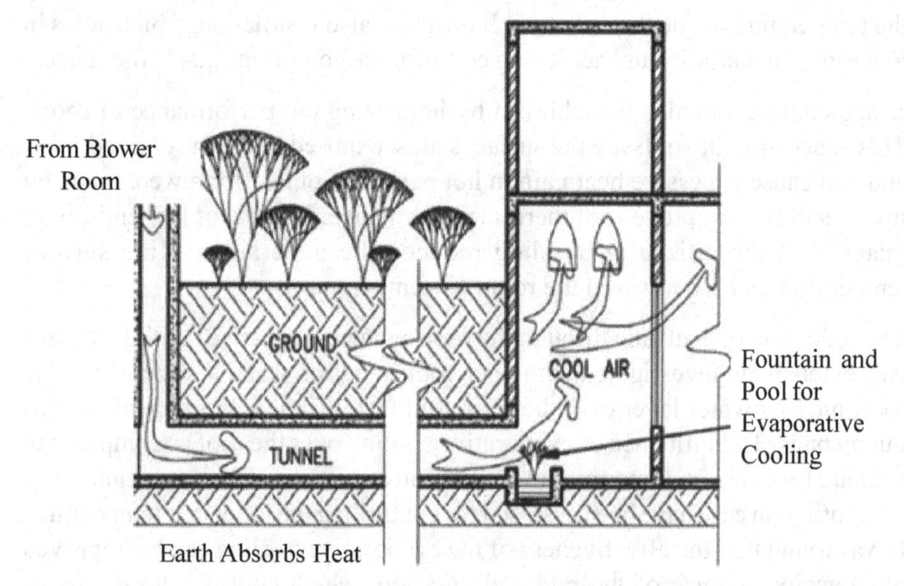

Fig. 3.4: Working principle of earth air tunnel

Earth berming: In an earth sheltered building or earth bermed structure the reduced infiltration of outside air and the additional thermal resistance of the surrounding earth considerably reduces the average thermal load. Further the addition of earth mass of the building acts like a large thermal mass and reduces the fluctuations in the thermal load. Besides reducing solar and convective heat gains, such buildings can also utilize the cooler sub-surface ground as a heat sink. Hence with reference to thermal comfort, an earth sheltered building presents a significant passive approach. Fig. 3.5 shows the working principle of earth berming during summer conditions.

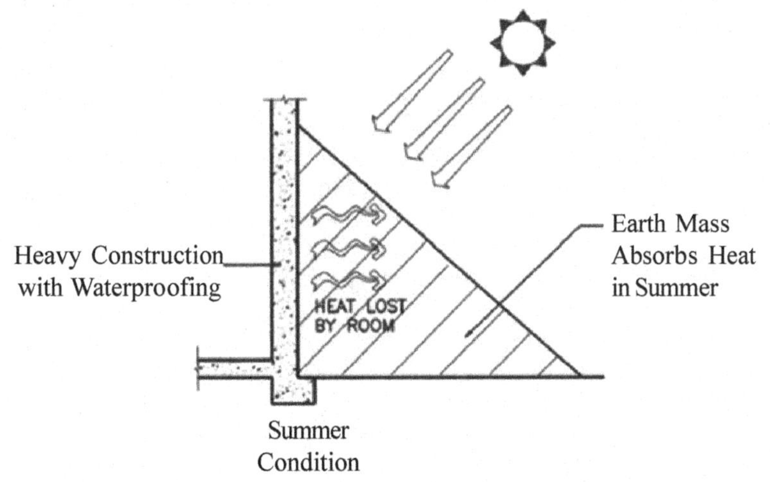

Fig. 3.5: Working principle of earth berming during summer

Summary

Passive cooling / heating techniques are moreover natural. These were so far developed specifically by understanding the conditions of location and weather. These are more cost and energy effective as well as environment friendly.

4

Theory of Solar Thermal Technology and its Application in Agriculture

N.M. Nahar

Chairman, Desert Science Institute, Jodhpur
and Former Principal Scientist, ICAR-CAZRI, Jodhpur, Rajasthan, India

Introduction

The age-old necessities of life are food, clothing and shelter. The 20th century had dramatized a fourth-energy. Energy starvation of the technological complex that maintains modern society may soon be as crucial a problem as feeding the hungry. Indeed, energy starvation could well precipitate more wide spread food starvation. Solutions for the energy crisis are strongly dependent on the technology of how energy is produced and used. To make a physical change in the world it is necessary to use four resources: energy, matter, space and time. How well a task has been performed can be measured in terms of amount of fuel consumed, the mass of material used, the space occupied, the hours of labour to accomplish it, and the ingenuity with which these resources are utilized. Squandering of irreplaceable energy resources, waste of materials, or large expenditure of space and time cannot be tolerated if the necessities of life are to be provided for all. Energy is intricately linked to development of a country. Growing demand on account of industrialization, urbanization, transportation and also increase consumption in rural areas is putting added pressure on energy supply network. Electricity generation has been increased from 1,400 MW in the year 1947 to 303,118 MW up to August 31, 2016 in India but still there is 10-15% shortage of power in peak hours and all villages are not electrified. Even many electrified villages receive 6 to 8 hours of electricity per day in India. To overcome this shortage there is a need to utilise new and renewable sources of energy. Fortunately, India is blessed with abundant solar energy. The arid parts of India receives much more radiation as compared to the rest of country i.e.

maximum solar radiation i.e. 7600-8000 MJm^{-2} (2111-2222 $kWhm^{-2}$) per annum, followed by semi-arid parts, 7200-7600 MJm^{-2} (2000-2111 $kWhm^{-2}$), per annum and least on hilly areas where solar radiation is still appreciable i.e. 6000 MJm^{-2} (1667 $kWhm^{-2}$) per annum (IMD 1985). In this chapter review of different solar thermal technologies in Indian perspective has been carried out.

Solar thermal technologies

Solar energy is absorbed by a black surface and transferred to the using medium by a suitable method. Thermal insulator is provided on the bottom of the black surface to reduce conduction heat loss and covered by a transparent sheet to reduce radiation heat loss. For low temperature applications i.e. below 100°C, flat plate collector is used while for higher temperatures concentrators are used. The solar energy can be efficiently used for cooking, water heating, drying, distillation, space heating, cooling, refrigeration and power generation in etc. in agriculture.

Solar water heaters

Hot water is an essential requirement in industries as well as in domestic sector. It is required for taking baths and for washing clothes, utensils and other domestic purposes both in urban as well as in rural areas. Hot water is required in large quantities in hostels, hotels, hospitals, industries such as textile, paper, food processing, dairy, edible oil etc. Water is generally heated by burning non-commercial fuels, namely firewood and cow dung cake in rural areas and by commercial fuels such as kerosene, liquid petroleum gas (LPG), coal, furnace oil and electricity in urban areas and in industries. Solar water heaters, therefore, seem to be a viable alternative to conventional fuels for water heating.

Natural circulation type solar water heater

The most commonly used solar water heater for domestic needs is natural circulation type. This type of solar water heater was designed, developed and investigated in detail by Close (1962), Yellot and Sobotaka (1964), Gupta and Garg (1968), Ong (1974), Nahar (1984), Morrison and Tran (1984), Morrison and Braun (1985), Vaxman and Sokolov (1986), Nahar and Gupta (1987), Norton et al. (1987), Nahar (1988, 1991, 1992, 2002, 2003) and Brinkworth (2001). Potential of solar water heater is 140 million m^2 but up to 31.12.2015 only 8.9 million m^2 have been installed. The programme of solar water heater has been initiated with subsidy but for the last 35 years only 6 % of actual potential have been installed. This is because flat plate collector manufactured in India is fabricated from copper tube and copper sheet that is expensive. Considering this Nahar (2002) at the Central Arid Zone Research Institute, Jodhpur has developed solar water heater that uses a flat plate collector made of galvanised steel (GS) tube and aluminium sheet (Fig. 4.1). The cost of this solar water heater is 30% less as compared to commercially available solar water heater in India, while performance is almost same of both solar water heaters.

Fig. 4.1: Solar water heater with GS-Al and Cu-Cu collector

Integrated collector storage solar water heater

Natural circulation type solar water heater has collector and storage tank in separate units, therefore, its cost is high and beyond the reach of rural people. Considering this, the cost has been reduced by combining both collector and storage tank in one unit and collector-cum-storage type solar water heater was designed, fabricated and tested. Such type of solar water heaters were studied by Tanishita (1970), Richards and Chinnery (1967), Garg (1975), Nahar (1983), and Nahar and Gupta (1988, 1989), Fairman et al. (2001), Tripangnostopoulos et al. (2002), Smyth et al. (2003), Souliotis and Tripanagnostopoulos (2004), and Madhlopa et al. (2006). These solar water heaters are simple in design, low cost, easy in operation and maintenance and easy to install. But life of this solar water heater was less than 8 years, therefore, it did not become popular in India. Considering this, integrated collector storage (ICS) solar water heater has been design, developed and fabricated. The life of this solar water heater is more than 15 years.

The ICS solar water heater consists of a rectangular tank $1000 \times 1000 \times 100$ mm^3, made from 3 mm thick mild steel plate. The absorber area is 1.0 m^2. The capacity of tank is 100 litres. The tank performs the dual function of absorbing solar radiation and storing heated water. It is encased in a galvanized steel (22 SWG) tray having dimension $1240 \times 1233 \times 270$ mm^3 with about 100 mm glass wool insulation at the bottom as well as on the sides. Two glass covers have been provided over it. The front surface of the tank is painted with black board paint. Inner surface of the tank was painted with anti-corrosive paint. An insulation cover is hinged over it so that the heater can be covered by it in the evening at 16.00h for getting hot water till next day morning. The heater works on push through systems. In urban areas, the inlet of the heater can be connected to water supply line through a gate valve. Hot water can be obtained by opening gate valve and collected through the outlet pipe. A funnel/bucket is provided for rural use where there is no water supply line. Hot water through outlet pipe can be obtained by putting cold water in the funnel/bucket. The heater is facing equator on a mild steel angle stand with >+15° tilt from horizontal for receiving maximum solar radiation during winter. Fig. 4.2 depicts actual installation of integrated collector storage solar water heater in the village Newara Road, Osian tehsil, Jodhpur.

Fig. 4.2: Integrated collector storage solar water heater at village Newara Road

Solar Dryer

In many rural areas of India, the farmers grow fruit and vegetables. These perishable commodities have to be sold in the market immediately after harvesting. When the production is high, the farmers have to sell the material at very low price, there by incurring great loss. This loss can be minimised by dehydrating fruits and vegetables. The dried products can be stored for longer time in less volume. In off seasons the farmer can sell the dried products at higher price. The traditional methods for drying the agricultural produce is to dehydrate the material under direct sunshine. This method of drying is a slow process and usual problems like dust contamination, insect infestation and spoilage due to unexpected rain. These problems can be solved by using either oil-fired or gas fired or electrically operated dryers. However, in many rural areas in India, the electricity is either not available or too expensive for drying purpose. Thus in such areas the drying systems based on the electrical heating are inappropriate. Alternatively, fossil powered dryer can be used but it poses such financial barriers due to large initial and running cost that these are beyond the reach of small and marginal farmers. In the present energy crisis, it is desirable to apply a little solar technology for dehydration of fruits and vegetables, so that gas, oil and electricity can be saved. India is blessed with abundant solar energy, which can be used for dehydrating fruits and vegetables through solar dryer. Keeping this in view, solar dryers both direct type viz. simple solar cabinet dryer, improved dryer with chimney, dryer for maximum energy capture, multirack tilted type dryer, and forced circulation type dryer have been designed, developed and tested at Central Arid Zone Research Institute (CAZRI), Jodhpur (Thanvi and Pande 1989). The details of the large size solar dryer are described below.

Large size solar dryer

A large size solar dryer (Fig. 4.3) that can be commercially used for drying fruits and vegetables was developed (Thanvi 1994). The salient features of this dryer are:

(i) It can capture the maximum solar energy throughout the year by keeping the system at optimum tilt during different seasons.

(ii) It can protect the drying material from rain, flies and squirrel.

(iii) Stainless steel wire mesh is used for fabrication of drying trays.

(iv) Partitions are provided in the drying trays so that the material can be stacked even on inclined plane.

(v) A low cost material viz. bajra stem or husk is used as insulation.

(vi) The dryers can be connected in series and hence its capacity can be enhanced as per requirement.

(vii) It can be dismantled easily so that its transportation is easy from one place to another.

Drying trials for dehydrating vegetables *viz.* mint, spinach, okra, tomato, ginger, red and green chillies, carrot, coriander leaves, fenugreek, peas, cabbage, onion, sweet potato, bitter gourd, radish, sugar beet, cauliflower, bathua and fruits, *viz.* ber, sapodilla, grapes, pomegranate, etc. were conducted successfully. The leafy vegetables can be dehydrated within 2 to 3 days at the loading rate of 4 to 5 kg/m^2, whereas other vegetables can be dried within 3-4 days at loading rate of 8 to 10 kg/m^2. Thus in general, it can be concluded that in commercial solar (glass area 10 m^2) about 100 kg of vegetables can be dried in 4 days. The green colour and aroma of solar dried products remained as such even after drying. These solar dried vegetables should be soaked in hot water before cooking. The spinach powder can be used for making 'Palak paneer'. The coriander and tomato powder can be mixed with ingredients to prepare instant soup/sauce/chutney by adding water. The solar dried grated carrot can be used for preparing pudding 'Gajar ka Halwa'.

Advantages of solar dryers

(i) Solar dryer can save fuel and electricity as required in case of mechanical drying method.

(ii) Drying time in solar dryer is reduced in comparison to open sun drying method.

(iii) Fruits and vegetables dried in solar dryer are better in quality and hygienic than dried in open.

Fig. 4.3: Large size solar dryer at village Keru, Jodhpur

(iv) The limited space available in houses in large cities can be effectively used for dehydrating fruits and vegetables using domestic solar dryer.

(v) Materials required for fabrication of solar dryer are locally available.

(vi) Use of solar dryer involves no fire risks.

(vii) The trade of dried vegetables can be linked with national and international trades.

Domestic Solar Cookers

Cooking accounts for a major share of energy consumption in developing countries. Most of the cooking requirement is met by non-commercial fuels such as firewood, agricultural waste and animal dung cake in rural areas and cooking gas and kerosene in urban areas in India. The fuel wood requirement is 0.4 ton per person per year in India. In rural areas fire wood crisis is far graver than that caused by a rise in oil prices. Small and marginal villagers have to forage 8-10 hour a day in search of fire wood as compared to 1-2 hour ten years ago. One third of India's fertilizer consumption can be met if animal dung is not burnt for cooking and is used instead as manure. The cutting of firewood causes deforestation that leads to desertification. Therefore, solar cookers seem to be good substitute for conventional cooking.

There are three broad categories of solar cookers (i) Reflector/focussing type (ii) Heat transfer type (iii) Hot box type.

The reflector type solar cooker was developed in early 1950's (Ghai 1953) and was manufactured on a large scale in India (Ghai et al. 1953). Attempts were also made is 1960's and 70's to develop a reflector type solar cooker (Duffie,

1961; Lof and Fester 1961; Tabor 1966; Von Oppen 1977). However a reflector type solar cooker did not become popular due to its inherent defects e.g. it required tracking towards sun every ten minutes, cooking could be done only in the middle of the day and only in direct sunlight, its performance was greatly affected by dust and wind, there was a danger of the cook being burned as it was necessary to stand very close to the cooker when cooking and the design was complicated.

In the heat transfer type solar cooker, collector is kept outside and cooking chamber is kept inside kitchen of the house (Abot 1939; Alward 1972; Garg and Thanvi 1977). But this type of solar cooker also did not become popular because of its high cost and only limited cooking can be performed.

The third type of cooker is known as hot box in which most of the defects of above two types of cookers have been removed (Ghosh 1956; Telkes 1959; Garg 1976). Different types of solar cookers have been tested and solar oven (Garg et al. 1978; Malhotra et al. 1982, 1983; Nahar 1986; Olwi and Khalifa, 1988) has been found best. Though performance of solar oven is very good but it also requires tracking towards sun every 30 minutes, it is too bulky and is costly. Therefore, the hot box solar cooker with single reflector (Parikh and Parikh 1978) is therefore being promoted by Ministry of New and Renewable Energy, Govt. of India and state nodal agencies and 1.33 million number solar cookers have been sold all over India up to December 31, 2016 (MNRE 2016). The performance of hot box solar cooker is very good during summer but it is very poor during winter in the northern parts of India because its absorbing surface is horizontal, and solar radiation received by a horizontal surface is 33% less as compared to a tilted surface in the winter season.

Therefore considering this a new solar cooker (Nahar 1990) has been designed which is having tilted absorbing (TA) surface and does not require frequent tracking but it had special cooking utensils therefore there is a difficulty in handling cooking material, therefore double reflector solar cooker with TIM has been developed (Nahar 2001) that does not require any tracking for three hours and performance is better in winter season because TIM reduces convective heat losses to a great extent (Nahar et al. 1994). To eliminate tracking completely a non-tracking solar cooker (Fig.4.3) was developed by Nahar (1998).

Fig. 4.3: Non-tracking solar cooker

Community solar cooker

Fig. 4.4: Community solar cooker in the field

A Community solar cooker (Fig. 4.4) capable of cooking for about 80 persons have been designed fabricated and tested (Nahar et al. 1993). The cooker is suitable for hostels, temples, canteens, restaurants, etc. The cooker can be used for boiling, roasting and baking. All dishes can be prepared within 2 to 3 hours. The cooker has been designed by considering length to breath ratio of reflector 1:4 so that it does not require any tracking. The cooker saves 20,700 MJ of energy per year. This cooker is suitable for cooking mid day meal in schools of rural areas.

Solar cooker for animal feed

During the survey of rural arid areas, it was found that huge amount of firewood, cow dung cake and agriculture wastes are burnt for boiling of animal feed. The solar cookers available are costly and cook only 2 kg of animal feed per day. Therefore, it was felt that a very low cost suitable solar cooker should be designed for boiling of animal feed. Considering this, a small simple novel solar cooker using locally available materials e.g. clay, pearl millet husk and animal dung have been designed, developed and tested that can boil 10 kg of animal feed per day.

A pit of dimensions 1980 mm × 760 mm × 100 mm is dug in the ground. The clay, pearl millet husk and animal dung have been mixed in equal proportion with water to make paste. The bottom of the pit has been filled up to a depth of 50 mm by this paste. It was left to dry in air for couple of days. The sides of solar cooker have been made by the same material, 150 mm pearl millet husk insulation has been provided at the bottom. 24 SWG galvanised steel absorber has been put over the insulation. The absorber has been painted with black board paint. Two glass covers (4mm thick) on a removable angle iron and wooden frame have been provided over it. Technician's help has been taken for fixing glass sheet on a suitable wooden frame. Four aluminium pans (Common name tagari) with lid can be kept inside cooking chamber for boiling of animal feed. The body of the cooker has been fabricated by an unskilled village labour. The solar cooker for animal feed can also be constructed from brick/stone masonry or vermiculite-cement. Crushed barley (*Jau Ghat*), guar korma, and gram churi with water were kept at 9:00h and were successfully boiled by

16:00 h. Solar cooker for animal feed is suitable for boiling 10 kg of food per day. Cooker saves 3671 MJ of energy per year. Low cost solar cookers for animal feed were installed at village Newara Road, Osian tehsil of Jodhpur. The body of the cookers has been fabricated by an unskilled labour. The body of the two solar cookers (Fig. 4.5) were made from cement concrete as desired and made by the farmers cost. Animal feed viz. cotton seed and khal have been successfully boiled by the farmers.

Fig. 4.5: Animal feed solar cooker at village Newra Road, Jodhpur

The solar cooker saves time of rural women and 1058 kg of fuel wood per year. It is easy to fabricate at village level and village carpenter will get job for the fabrication of glass frame which is also very simple. The use of solar cooker for animal feed would help in conservation of conventional fuels, such as firewood, cow dung cake and agricultural waste in rural areas of India. Conservation of firewood help in preserving the ecosystems and cow dung cake could be used as fertiliser, which could aid in the increase of production of agricultural products. Moreover, the use of the solar cooker for animal feed would result in the reduction of the release of CO_2 to the environment and getting CER under CDM mechanism of UNFCCC.

Solar Stills

In arid zone of India, there is acute shortage of drinkable water. Generally in summer season, villagers travel many miles in search of fresh water. It is observed that at least two or three family members are always busy in bringing fresh water from distant sources. The worst conditions are generated if the resources of water are not available and villagers are forced to take highly saline underground water containing nitrate and fluorides or pond water contaminated with pathogenic microbes. This normally leads to cause the physical disorder of various kinds. It is further noticed that these areas are endowed with plentiful of solar energy which can be used for converting brackish water to demineralised water through solar stills. Two types of solar stills, *viz.* low glass roof type in masonry construction for desalination of brackish water on large scale and multiple basin type for providing distilled water for batteries have been designed, developed and tested at CAZRI, Jodhpur (Thanvi 1982).

Low glass roof type solar stills in masonry construction

Low glass roof type solar stills have been developed for desalination of brackish water. Structural improvements have been made in existing designs e.g. slope in distillation channel, supporting truss for ridge, windows for cleaning on two sides, better sealing, optimum glass slope of 15°C, square geometrical configuration to minimise peripheral heat losses etc. Thermal efficiency of uninsulated stills was found to range from 20 to 34% from winter to summer with productivity ranging from 1.0 to 3.3 litres m^{-2} day^{-1}. Efforts have also been made to mitigate the problems of algae and scale formations. The size of solar still can be matched as per requirement. The distillate output of solar still is to be mixed with the available saline water in appropriate proportion to make it drinkable (Thanvi 1996).

Multi basin tilted type solar still

The solar distillation technique has been found more useful for production of distilled water required for maintenance of batteries which are used in tractors, trains, aeroplanes, electric grid stations etc. The main advantage of this solar distillation in comparison to other conventional process are (a) the solar still is simple to fabricate and it will operate for long period with little attention (b) size of solar still can be matched with the demand of distilled water (c) The salinity of raw water may range from sea water to slightly brackish (d) No power or fuel is required (e) No extra water is required for condensing vapours (f) It is the only process which is practicable for production of distilled water even in isolated areas.

The Central Arid Zone Research Institute (CAZRI) Jodhpur has developed an efficient multi basin tilted type solar still (Thanvi 1982) for production of distilled water. Fig. 4.6 depicts installation of multi basin tilted type solar still in the field. This unit has the following features:

Fig. 4.6: Multi basin tilted type solar still

(i) Four aluminium trays are fixed in stepped fashion inside a wooden box having a glass cover at the top.

(ii) Fibreglass insulation is provided at the base of the trays.

(iii) An adjustable M.S. angle iron stand provided to keep the system at an optimum tilt in accordance with latitude and season of operation.

(iv) The saline water can be filled in each tray with the help of G.I. tube fixed to the top most tray.

(v) Water level is maintained by an overflow arrangement provided at the bottom tray.

(vi) For collecting the distillate, an aluminium channel is provided.

(vii) It was found that under Jodhpur conditions, a multi basin tilted type solar still supplies distilled water 3.1 to 4.1 litres/m^2 day throughout the year indicating that the distillate output of this novel still is not much affected by the seasons. The distillate output during winter is three to four times more compared to that obtained from conventional uninsulated single basin solar still.

Solar still for production of rose water

A farmer having irrigation facilities can grow the roses in his field. These roses can be fed in a specially designed tilted type solar still for production of rose water (Thanvi and Pande1981). This solar still has the following features:

(i) Specially designed stepped basin for keeping the rose petals and water conveniently even on inclined position

(ii) Openable cover which facilitates the cleaning operation. This unit (glass area 0.6 m^2) supplies 3.7 litres of rose water in 3 days during winter.

Dual and Multi-Purpose Solar Energy Devices

Improved multipurpose solar energy device

A novel device has been developed (Nahar et al. 1986) which can provide 80 litres of hot water at 50-60°C and 1 to 2 litres of distilled water per day. The same device can be used as a solar cabinet dryer as and when required (Fig. 4.7).

Fig. 4.7: Improved multipurpose solar energy device

Improved solar water heater-cum-steam cooker

A two-in-one device has been designed, fabricated and tested (Nahar 1985). The device can be used as a solar water heater during winter when it can provide 100 litres of hot water at 60-70°C in the evening which can be retained to 50-60° C till next day. The same device can be used as a steam cooker during summer for boiling 1 kg. of food per day.

Solar water heater cum solar cooker.

The design of collector-cum-storage type solar water heater has been modified so that it can be used as solar cooker (TA) during summer when hot water is not required (Nahar 1988)

Large size solar water heater cum solar cooker

A novel device has been designed, fabricated and tested (Nahar 1993). It can provide 150 litres of hot water at 50-60° C in the evening which can be retained to 40-45°C till next day. With minor adjustment it can be used as a hot box solar cooker for cooking food for about 40 people. The efficiency of the device as a water heater and as a cooker has been found to be 67.7% and 29.8% respectively.

Solar thermal power plant

Solar power plants are based on either photovoltaic conversion or thermal conversion. Solar thermal power plants based on flat-plate collector are having efficiency only 1 to 2%, therefore, solar thermal power plants based on solar concentrators are only described below.

A solar thermal power plant comprises of a solar concentrator for concentrating solar energy on to receiver where suitable heat transfer fluid is heated. This fluid transfers the required heat to the working medium through a heat exchanger. Excess heat is transferred to thermal storage system. Working medium is usually steam, which is heated to high temperature at high pressure. There could be a few heat to mechanical conversion cycles (Ranking cycle is most popular). Obviously, heart of the device is this solar concentrator. On the basis of design of solar concentrator solar thermal power plants could be divided into four categories, which are as follows:

(1) Central receiver system (Vant-Hull and Easton 1975, Vant-Hull and Hilderbrandt 1982 and Wehowsky and Stahl 1983), (2) Parabolic dish system (Simpson et al. 1982), (3) Hemispherical bowl system (Simpson et al. 1982), (4) Parabolic trough system (Ann. 1985)

Central Receiver System

In a central receiver system, a field of sun tracking mirrors (heliostats) reflects sunlight into a receiver located on a tower, heating a fluid that circulates within the receiver. The principal advantage of the central receiver system is their ability to collect energy at high temperatures and favourable costs. A central receiver system has five main components. (1) Heliostats (2) a receiver (3) heat transport and exchange system (4) thermal storage (optional) and (5) controls. The heliostats track the sun automatically during the day, reflecting the sun's radiant energy onto the receiver. The concentrated energy is absorbed efficiently by a fluid circulating within the receiver. The resulting thermal energy, which can be collected at temperatures up to 1100°C, can be used directly or indirectly to drive a turbine to produce electrical power, as process heat in industrial applications, or to produce fuels and chemicals. If a storage sub-system is incorporated, energy can be stored for use when the sun is not shining. Automatic operation and sub-system interactions are accomplished by a master control system. Ten solar power plants based on tower system have been commissioned at different places e.g. Barstow, Southern California, USA (10 MW), Almeria, Spain (500 kW, 1.2 MW), Nio-Cho Kajawa, Japan (1.0 MW), Eurolies EFC (1.0 MW), Thesis, France (2.5 MW), Sandia National Laboratories, USA (5 MW), SES-5, USSR (5.0 MW), PHOEBUS-TSA, Spain (2.5 MW) and Solar two, USA (10 MW) and Bikaner, India 2.5 MW.

Advantages of the central receiver concept

High solar efficiencies , high steam temperatures, simple hybridisation with fuel oil or natural gas, high power availability of more than 94%, modular solar components (heliostats) with high mass production potential, simple operation strategy, Process steam generation for eventual cogeneration.

Disadvantages of the central receiver concept

Thermal energy losses and parasitic (blower) due to the open air cycle (PHOEBUS), the solar energy and the fossil backup fuel are converted to electricity with a relatively low steam cycle efficiency, the heliostats require very stable support for the mirrors and two-axis tracking, water needed for mirror cleaning (alternatives are available).

Parabolic Dish Systems

The dish system concept consists of a dish-shaped parabolic concentrator that focuses the sun's rays onto a receiver (or absorber) mounted above the dish at its focal point. Sunlight is focused on an opening in the receiver to heat a fluid circulating within coils. The hot fluid can be transported elsewhere for a variety of thermal uses, or direct electric generation can be accomplished by integrating an engine or alternator with the receiver at the focal point. Each dish is a complete power-production unit (module) that can function either as an independent system or as part of a group of modules. A single large parabolic dish module can achieve a fluid temperature of up to 1500° C and can efficiently produce up to 50 kW of electricity, or 150 kW thermal. About sixteen types of parabolic dish systems have been designed and commissioned. The most important are Lajpat, California, USA (4.9 MW), Messerschmitt Bolkow Blohm (MBB) solar plant at Kuwait (100 kW), Shenandoahs, Georgia, USA (100-400 kW) and Huntington Beach test, USA (25 kW).

Advantage of parabolic dish system

Completely auto contained stand-alone unit, very high concentration ratios, working temperatures and efficiencies, long term experience with small-scale power plants and single units, options for distributed as well as centralised electricity supply systems, modularity of the system, benefits of mass production, no scale restriction., Simple operation and maintenance

Disadvantages of the dish-Sterling concept

Integrated fossil back up not available until now, low power availability and few annual full load hours, requires rigid support structures and perfect tracking that leads to high costs, no experience with large utility-scale systems, and water requirement for cleaning.

Hemispherical Bowl System

The solar bowl concept involves a fixed mirror collector, shaped in the form of a hemisphere and tilted upward from the horizon along a north-south line depending on the latitude of location. The slender cylindrical receiver, helical coiled tubing on pipe, is supported by a cantilevered beam pivoted on a dual-axis mount to track the sun. The system can convert pressurised water to superheated steam that can drive a turbine for the production of electricity. It can also use other heat-transfer fluids at a lower pressure and temperature than water for industrial process heat or for heating and cooling buildings. A hemispherical bowl project was developed by the Crosbyton (Texas) Solar Power project funded by the U.S. Department of Energy. Bowls can be larger than other single-unit concentration devices since the reflecting surface does not have to be moved to track the sun. A 65 ft. Diameter solar bowl was designed and constructed in the United States to determine the technical feasibility of the concept. This small scale system (18 ft. Boiler, 3300 ft^2 mirror surface 75 kW thermal) has been operational since January 1980 for more than 10,000 hours.

A preliminary design of a 200 ft diameter bowl system with 31,000 ft^2 of reflective collector surface has been completed. Using a 58.5 ft. Receiver made up of an 18ft. Diameter support pipe wrapped with 20 boiler tubes, this large-scale bowl system was designed to produce 500 kW at peak power. Areas of bowl technology presently being addressing include design of spherical curved mirrors, single and multiple bowl controls, system models, air flow patterns of multiple bowl arrays, mirror panel support structures, a mirror cleaning system, and solar receiver/boiler design and manufactures.

Parabolic trough system

A parabolic trough system is comprised of a field of trough shaped concentrators that focus sunlight onto a linear receiver tube positioned along the focal line of the trough. A fluid in the receiver is heated and is then transported to the point of use by means of an insulated piping network. Concentrating collectors use reflective surfaces to focus the sun's rays onto a receiver or absorber where the solar energy heats a circulating fluid. The hot fluid can then be used directly for an industrial process, to power a turbine for mechanical work, or to generate electricity. The trough usually rotates about one axis to follow (track) the sun. Thermal storage systems accumulate thermal energy to be used during cloudy weather or at night. Currently storage system capacities range from "buffer" storage for short time intervals, such as during cloud passage, to as long as six continuous hours. Modularity is an advantage of a parabolic trough system. The basic collector module is a row of troughs coupled to a drive motor that rotates the trough module about a single axis to track the sun's position. A control system connects as many modules as required to raise the fluid in the troughs to a specified outlet temperature. These sub-system are called delta-T loops and an array of such loops is called a field. A field can consist of one loop or many, depending upon the energy required for a particular application. The single-axis tracking trough concept exhibits favourable performance and cost-effectiveness in the mid-temperature range; i.e. 100°C to 350°C. Energy requirements within this range are significant, including industrial process heat, production of mechanical or electrical energy such as for irrigation pumping, steam generation for enhanced oil recovery, and "total energy" production or cogeneration (e.g. providing for both electrical and direct heating processes). These plants were manufactured by Acurex, MAN and LUZ. LUZ had developed solar power plants in two rated capacities of 13.8 MW and 30 MW. Later on LUZ was taken by another company and SEGS VIII and SEGS IX of 80 MW rated capacities were commissioned. More than 10 test and demonstration facilities had been erected since 1980s and 354 MW of SEGS have been installed. 50 MW solar thermal power plant by M/S Godavari Power and Ispat and 100 MW Reliance Power has been installed in 2013-14 in the Jaisalmer district, Rajasthan.

Advantages of parabolic trough system

Reliable technology with more than 5000 GWh operating experience with the oil cooled parabolic trough collectors, simple hybridisation with fuel oil or natural gas, high power availability of more than 94%, modulate solar components with high mass production potential, simple operation strategy, cogeneration is possible.

Disadvantages of parabolic trough system

Solar operating temperature restricted to 400° C, steam temperature restricted to 370° C in solar mode (510° C in fuel mode), the solar energy and the fossil backup energy are converted with a relative low steam cycle efficiency, the collectors require very stable supports for the mirrors, considerable thermal inertia of the HTF loop, high cosine losses due to one-axis tracking, water needed for cooling and cleaning.

Criteria for establishing solar thermal power plant

(i) **Solar Radiation:** The location should be plentiful in solar radiation with at least 300 clear days in a year. In this respect, Jodhpur is an ideal location. It receives 6.0 KW/m^2 day average solar radiation on horizontal surface and 6.27 KW/m^2 day average solar radiation at normal incidence. The average daily duration of sunshine hours are 8.9 hours (Gupta and Nahar 1994).

(ii) **Desired quantity of water:** Appreciable quantity of water is required both as a working fluid in the steam turbine as well as for cooling cycle. The estimated requirement of water for establishing 30 MW solar thermal power plant is about 100 million gallons per annum or approximately 0.3 million per day. The proposed site is near village Mathania, 30km far from Jodhpur. Water is available at a depth of about 30 metres and discharge of well is estimated to be around 10,000 to 20,000 gallons per hour. It is estimated that 3 to 4 well will be sufficient to provide enough water for this plant. The quality of water is also good. The electrical conductivity is around 2 to 4 millimhos. However, softening plant can be erected.

(iii) **Requirement of land:** Solar thermal power plants require more land as compared to conventional power plants. For solar power plant based on parabolic trough concentrators the requirements of land for establishing 30 MW power plant is about 200 acres. At the proposed site more than 200 acres of waste flatland is available with district revenue authorities. Of course suitable soil testing will be required from the soil mechanics point of view for the foundation of parabolic trough concentrators and for power block.

(iv) **Vicinity of user:** There is very good industrial area coming up at the immediate vicinity of proposed site. Industries will be eager to utilise the generated power. Some of the industries could make use of low grade thermal heat from the heat exchanger in the cooling cycle and therefore, principle of concentration could be adopted to produce over-all power both electric and thermal at lower cost.

(v) **Vicinity of central grid:** Central grid of 132 kV is likely to be passed within a few kms of the site and therefore, transmission losses will also be minimum.

(vi) **Accessibility of location:** The proposed site is on the Jodhpur Phalodi main metalled road and Jodhpur Jaisalmer railway line (Broad gauge), therefore, transportation of necessary material to be site is likely to be very easy. Besides Jodhpur is well connected with Delhi-Bombay Air and train routes, therefore, visits of foreign/Indian experts will be comfortable.

Potential of solar thermal power plant in the Thar Desert

The Indian hot arid zone occupies an area of 3,20,000 km^2 spread over western Rajasthan, north Gujarat, south west of Haryana and Punjab and parts of Andhra Pradesh and Karnataka. But the major part of the Indian arid zone (62%) lies in the western part of the Rajasthan covering 12 districts. Table 4.1 shows global, diffuse and direct solar radiation received at Jodhpur, the gateway of the Thar Desert and Table 4.2 shows solar radiation received on different parts of arid regions. If 2% arid area is covered under solar thermal power plant than 320,000 MW power can be generated while present installed capacity is only 303,118 MW in India.

Table 4.1: Solar radiation at Jodhpur (KWh M^{-2} day^{-1})

Month	Global	Diffuse	Direct	Sunshine (hrs.)
January	4.715	1.138	7.238	9.1
February	5.565	1.362	7.420	9.1
March	6.540	1.771	7.354	9.0
April	7.233	2.336	6.725	9.5
May	7.545	2.683	6.324	9.8
June	7.068	3.058	4.985	9.2
July	5.979	3.385	3.298	6.9
August	5.544	3.208	3.126	6.8
September	6.101	1.878	6.020	9.4
October	5.827	1.218	7.604	9.7
November	4.903	0.808	7.719	9.4
December	4.432	0.931	7.448	9.0
Average	5.954	1.981	6.272	8.9

Table 4.2: Global solar radiation (KWh M^{-2} day^{-1}) at different arid stations of Rajasthan, Gujarat and Haryana

State and stations	Winter (Dec.-Feb.)	Summer (March -May)	Monsoon (Jun - Sep.)	Post monsoon (Oct.-Nov.)	Annual average
Rajasthan					
Jodhpur	4.92	6.93	6.43	5.52	5.92
Bikaner	4.67	7.33	7.27	5.38	6.16
Jaisalmer	4.84	7.36	7.39	5.52	6.27
Barmer	4.92	7.36	6.82	5.57	6.16
Hanumangarh	4.20	6.33	6.10	5.60	5.55
Gujarat					
Bhuj	5.19	6.85	5.73	5.65	5.85
Haryana					
Hisar	4.28	6.52	6.02	5.76	5.64

Constraints

Biggest advantage of solar thermal power plant is that it can be hybridised with conventional fuel and it can be operated round the clock in absence of solar radiation. Solar thermal power plant at California are hybridised with natural gas therefore these plants are successfully in operation from 1990 to date.

But policy of Ministry of New and Renewable Energy, Govt of India is that these should be operated standalone that is only on solar radiation. Few solar thermal power plant of total 152.5 MW capacity have been installed in the Thar Desert in Rajasthan but these are facing staring problems because for starting next day these need auxiliary energy and unable to operate more than 6 hours since these depend only on availability of direct solar radiation.

Therefore there is need to change the solar thermal power plant policy under National solar energy mission 2022 so that these plants can run smoothly if hybridised with natural gas/ other fuel in the ratio 60:40 (solar : conventional fuel) load factor as SEG I to SEG IX Plants in California.

Conclusions

Solar energy devices like solar water heater, solar dryer, solar cooker, solar cooker for animal feed, solar still etc. are very much useful in rural areas and hence would help in rural development. Bio-gas and smokeless stoves will meet cooking fuel requirement in rural areas. The use of these devices would help in conservation of conventional fuels, such as firewood, cow dung cake and agricultural waste in rural areas of India. Conservation of firewood help in preserving the ecosystems and cow dung cake could be used as fertiliser, which could aid in the increase of production of agricultural products. Moreover, the use of these devices would result in the reduction of the release of CO_2 to the environment and getting CER under CDM mechanism of UNFCCC.

Solar thermal power plant based on the parabolic trough system is most suitable power plant for Indian conditions. It produces cheapest electricity among all solar power plants. It is easy in operation and can easily be hybridised with the fossil fuel back up system. If 2.0% arid area is covered under solar thermal power plant than 320,000 MW power can be generated while present installed capacity is only 303,118 MW in India.

References

Abot, C.G. 1939. Smithsonian Misc. Cells, 98.

Alward, R.1972. Solar steam cooker Do it yourself leaflet L-2 Brace Research Institute, Quebec, Canada.

Ann. 1985. Solar thermal technology annual evaluation report fiscal year 1984. US Department of Energy, DOE/CE/T-13.

Brinkworth, B. J. 2001. Solar DHW performance correlation revisited. Solar Energy 71 :389-401.

Close, D.J. 1962. The performance of solar water heater with natural circulation. Solar Energy 6 (1)33-40.

Duffie, J.A., Lof, G. O. G. and Beck, B. 1961. Laboratories and field studies of plastic reflector solar cookers, Proceedings of UN Conference on New sources of Energy, Rome, Paper S/87, 5, 339-346.

Fairman D., Hazan, H. and Laufer, I. 2001. Reducing heat loss at night from solar water heaters of the integrated collector storage variety. Solar Energy 71: 87-93.

Garg, H.P.1976. A solar oven for cooking. Indian Farming 27: 7-9.

Garg H.P. and Thanvi K.P. 1977. Studies on solar steam cooker. Indian Farming 27 (1): 23-24.

Garg, H.P., Mann, H.S. and Thanvi K.P. 1978. Performance evaluation of fire solar cookers, Proc. ISES Congress, New Delhi (Eds. F de Winter and M. Cox) Pergamon Press New York SUN 2, 1491-1496.

Garg, H.P. 1975. Year round performance studies on a built-in-storage type solar water heater at Jodhpur. Solar Energy 17: 167-172.

Ghai, M.L. 1953. Design of reflector type direct solar cooker. Journal of Scientific and Industrial Research 12 A: 165-175.

Ghai, M.L., Pandhar, B.S. and Dass H. 1953. Manufacture of reflector type direct solar cooker. Journal of Scientific and Industrial Research 12A: 212-216.

Ghosh, M.K. 1956. Utilisation of solar energy. Science and Culture 22: 304-312.

Gupta, C.L. and Garg, H.P. 1968. System design in solar water heaters with natural circulation Solar Energy 12: 163- 182.

Gupta, J.P. and Nahar, N.M. 1988. Feasibility of 30 MW solar thermal power stations at Jodhpur. The Indian Journal of Energy 23:468-470.

Gupta, J.P., Nahar, N.M., Purohit, M.M. and Pushpak, S.N. 1994. Deterioration in specular reflectance of mirror materials by exposure to desert environment. Proc. National Solar Energy Convention, Gujarat Energy Development Agency Vadodara., pp.42-44.

IMD. 1985. Solar radiation atlas of India, India Meteorological Department, New Delhi.

Lof, G.O.G. and Fester, D. A. 1961. Design and performance of folding umbrella type solar cooker Proceedings of U N Conference on new sources of Energy, Rome, Paper S/100, 5: 347-352.

Madhlopa A., Magawi, R. and Taulo, J. 2006. Experimental study of temperature stratification in an integrated collector-storage solar water heater with two horizontal tanks. Solar Energy 80: 989-1002.

Malhotra, K.S., Nahar, N.M., and Ramana Rao, B.V. 1982. An improved solar cooker. Internatinal Journal of Energy Research 6 : 129-33.

Malhotra, K.S, Nahar, N.M. and Ramana Rao B.V. 1983. Optimization factor of solar ovens. Solar Energy 31: 235-237.

MNRE (2015). Annual Report. Ministry of New and Renewable Energy, Government of India, New Delhi.

Morrison G. L. and Tran, N. H. 1984. Simulation of the long term performance of thermosyphon solar water heaters. Solar Energy 33: 515-26.

Morrison G. L. and Braun, J. E. 1985. System modelling and operation characteristics of thermosyphon solar water heater. Solar Energy 34: 389-405. '

Nahar, N.M. 1983. Year round performance of an improved collector-cum-storage type solar water heater. Energy Conversion and Management 23: 91-95.

Nahar, N.M. 1984. Energy conservation and field performance of natural circulation type solar water heater. Energy 9: 461-464.

Nahar, N.M. 1985. Performance and testing of an improved solar water heater cum steam cooker. International Journal of Energy Research 9:113-116.

Nahar, N.M. 1986. Performance studies on different models of solar cookers in arid zone conditions of India. Proc. 7th Miami International Conf. on Alternative Energy Sources Hemisphere Publishing Corporation, New York 1:431-439.

Nahar, N.M., Thanvi, K.P. and Raman Rao, B.V. 1986. Design, development and testing of an improved multipurpose solar energy device. International Journal of Energy Research 10:91-96.

Nahar, N.M. and Gupta, J.P. 1987. Performance and testing of improved natural circulation type solar water heater in arid areas. Energy Conversion and Management 27: 29-32.

Nahar, N.M. 1988. Performance and testing of a natural circulation solar water heating system. International Journal of Ambient Energy 9: 149-154.

Nahar, N.M. 1988. Performance and testing of a low cost solar water heater-cum-solar cooker. Solar and Wind Technology 5: 611-615.

Nahar, N.M. and Gupta, J.P. 1988. Studies on collector-cum-storage type solar water heaters under arid zone conditions of India. International Journal of Energy Research 12: 147-53.

Nahar N. M. and Gupta, J. P. 1989. Energy conservation and payback periods of collector-cum-storage type solar water heaters. Applied Energy 34: 155-162.

Nahar, N.M. 1990. Performance and testing of an improved hot box solar cooker. Energy Conversion and Management 30: 9-16.

Nahar N. M. 1991. Energy conservation and payback periods of large size solar water heater. Energy Conversion and Management 32: 371-374.

Nahar, N. M. 1992. Energy conservation and payback periods of natural circulation type solar water heaters; International Journal of Energy Research 16: 445-452.

Nahar, N.M. 1993. Performance and testing of large size solar water heater cum solar cooker International Journal of Energy Research 17: 57-67.

Nahar, N.M., Gupta, J.P. and Sharma, P. 1993. Performance and testing of an improved community size solar cooker. Energy Conversion and Management 34: 327-333.

Nahar, N.M., Gupta, J.P. and Sharma, P. 1994. Design, development and testing of a large size solar cooker for animal feed. Applied Energy 48: 295-304.

Nahar, N.M., Marshall, R.H. and Brinkworth, B.J. 1994. Studies on a hot box, solar cooker with transparent insulating materials. Energy Conversion and Management 35: 784-791.

Nahar, N. M. 1998. Design, development and testing of a novel non tracking solar cooker. International Journal of Energy Research 22: 1191-1198.

Nahar, N. M. 2001. Design, development and testing of a double reflector hot box solar cooker with a transparent insulation material. Renewable Energy 23: 167-79.

Nahar N. M. 2002. Capital cost and economic viability of thermosyphonic solar water heaters manufactured from alternate materials in India. International Journal of Renewable Energy 26: 623-635.

Nahar N. M. 2003. Year round performance and potential of a natural circulation type solar water heater. International Journal of Energy and Buildings 35: 239-247.

Norton B., Probert, S. D. and Gidney, J. T. 1987. Diurnal performance of thermosyphonic solar water heaters - an empirical prediction method. Solar Energy 39: 251-65.

Olwi, I.A. and Khalifa A.H. 1988. Computer simulation of solar pressure cooker. Solar Energy 40: 259-268.

Ong, K.S. 1974. A finite difference method to evaluate the thermal performance of a solar water heater. Solar Energy 16: 137-148.

Parikh, M. and Parikh, R. 1978. Design of flat plate solar cooker for rural applications. Proc National Solar Energy Convention of India, Bhavnagar, Central Salts and Marine Chemical Research Institute, Bhavnagar, India. pp. 257-261

Richards, S.J. and Chinnery, D.N.W. 1967. A solar water heater for low cost housing NBRI Bull. 41 CSIR RES Report South Africa 273, pp.1-26.

Simpson, T. L., O' Hair, E. and Reichert, J.D. 1982. The Crosbyton solar Power Project. Vol. VIII Preliminary design of 5 MW solar fossil Hybrid Electric Power Plant at Crosbyton, Texas, Texas Tech. University, Lubbock, TX 79409.

Souliotis M. and Tripanagnostopoulos, Y. 2004. Experimental study of CPC type ICS solar systems. Solar Energy 76: 389-408.

Smyth M. Eames, P.C. and Norton, B. 2003. Heat retaining integrated collector/storage solar water heaters. Solar Energy 75: 27-34.

Tabor, H. 1966. A solar cooker for developing countries. Solar Energy 10: 153-157.

Tanishita, I. 1970. Present situation of commercial solar water heater in Japan, Proc. of ISES Conf. Malbourne, Australia, Paper No.2/73.

Telkes, M. 1959. Solar cooking ovens. Solar Energy 3, 1-11.

Thanvi, K.P. and Pande, P.C. 1981. Designing a suitable solar still. Proc. National Solar Energy Convention, IISC, Bangalore (India), pp. 4020-4022.

Thanvi, K. P. 1982. Design, development of a multi basin tilted type solar still. Proc. National Solar Energy Convention. pp 7.001- 7.004.

Thanvi, K.P., and Pande, P.C. 1989. Performmance evaluation of solar dryers developed at CAZRI, Jodhpur. Solar Drying (Eds. A.N. Mathur, Yusuf Ali and R.C. Maheshwari), Himanshu Publication, Udaipur, pp. 158-164.

Thanvi, K. P. 1994. Studies on drying of okra in an inclined solar dryer. Proc. National Solar Energy Convention, Gujarat Energy Development Agency Vadodara. pp 77-80.

Thanvi K. P. 1996. Development of solar stills for production of drinking water in arid regions. Proc. National Seminar on New Strategies of Water Resources Management for 21st century. Department of Civil Engineering, JNV University, Jodhpur, pp. 233-241.

Tripanagnostopoulos Y., Souliotis, M. and Nousia, Th. 2002. CPC type integrated collector storage systems. Solar Energy 72: 327-350.

Vant-Hull, L.L. and Easton, C.R. 1975. Solar thermal power systems based on optical transmission (a feasibility study). Final report NSF/RANN/SE/GI-39456/FR/75/3. Pp. 410.

Vant-Hull, L.L. and Hilderbrandt, A.F. 1982. The 10 MW Central receiver pilot plant at Barstow, California Progress in Solar Energy. pp. 361-366.

Wehowsky, P. and Stahl, D. 1983. The gas cooled solar power project GAST. In: Solar World Congress, ISES Perth, Australia.

Von Oppen, M. 1977. The Sun basket. Appropriate Technology 4: 8-10.

Yellot, J.I. and Sobtka, R. 1964. An investigation of solar water heater performance. Trans. ASHRAE 70: 425-433

5

Fundamental Theories and Laws of Solar Radiation and Solar PV Technology

Priyabrata Santra and S. Poonia

ICAR-Central Arid Zone Research Institute, Jodhpur, Rajasthan, India

For electromagnetic radiation, there are four "laws" that describe the type and amount of energy being emitted by an object. In the following sections, these four laws of radiation e.g. Planck's law, Wien's displacement law, Steffan-Boltzman law and Kirchhoff's law are described briefly.

Planck's Law

Planck's law describes the spectral density of electromagnetic radiation emitted by a black body in thermal equilibrium at a given temperature *T*. The spectral radiance of a body, B_ν, describes the amount of energy it gives off as radiation of different frequencies. It is measured in terms of the power emitted per unit area of the body, per unit solid angle that the radiation is measured over, per unit frequency. Planck showed that the spectral radiance of a body at absolute temperature T is given by

$$B_\nu(\nu,T) = \frac{2h\nu 3}{c^2}\frac{1}{e^{\frac{h\nu}{kT}}-1}$$

where *k* is the Boltzmann constant, *h* the Planck constant, and *c* the speed of light in the medium, whether material or vacuum. The Boltzmann constant (k_B or *k*), named after Ludwig Boltzmann, is a physical constant relating energy at the individual particle level with temperature. It is the gas constant *R* divided by the Avogadro constant N_A:

$$k = \frac{R}{N_A}$$

The Boltzmann constant has the dimension of energy divided by temperature, the same as entropy. The accepted value in SI units is 1.38064852(79)×10^{-23} J/K^{-1}. The law may also be expressed in other terms, such as the number of photons emitted at a certain wavelength, or the energy density in a volume of radiation. The SI units of B_i are W·sr^{-1}·m^{-2}·Hz^{-1}.

In the limit of low frequencies (i.e. long wavelengths), Planck's law tends to the Rayleigh–Jeans law, while in the limit of high frequencies (i.e. small wavelengths) it tends to the Wien approximation (Fig. 5.1).

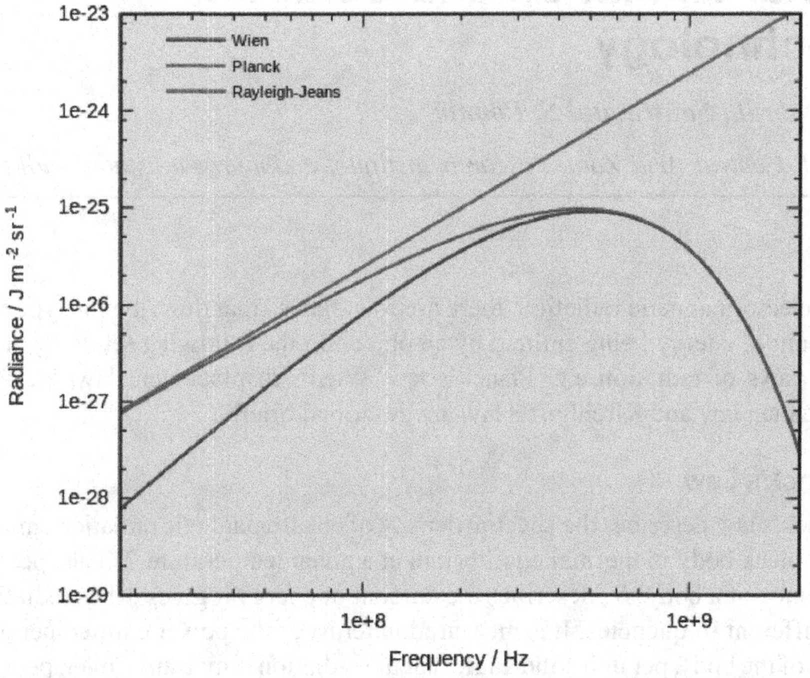

Fig. 5.1: Approximation of Planck's law with Rayleigh–Jeans law and Wien approximation

The Rayleigh–Jeans law attempts to describe the spectral radiance of electromagnetic radiation at all wavelengths from a black body at a given temperature through classical arguments. For wavelength λ, it is:

$$B_{\lambda}(T) = \frac{2ckT}{\lambda^4}$$

Wien's approximation (also sometimes called Wien's law or the Wien distribution law) is a law of physics used to describe the spectrum of thermal radiation (frequently called the blackbody function). This law was first derived by Wilhelm Wien in 1896. The equation does accurately describe the short wavelength (high frequency) spectrum of thermal emission from objects, but it fails to accurately fit the experimental data for long wavelengths (low frequency) emission. Wien derived this law from thermodynamic arguments, several years before Planck introduced the quantization of radiation and the law may be written as

$$I(\nu,T)=\frac{2h\nu^3}{c^2}e^{-\frac{h\nu}{kT}}$$

Where, *I*(n,*T*) is the amount of energy per unit surface area per unit time per unit solid angle per unit frequency emitted at a frequency í, *T* is the temperature of the black body, *h* is Planck's constant, *c* is the speed of light and *k* is Boltzmann's constant.

Wein's displacement Law

Wien's displacement law states that the black body radiation curve for different temperatures peaks at a wavelength inversely proportional to the temperature. The shift of that peak is a direct consequence of the Planck radiation law which describes the spectral brightness of black body radiation as a function of wavelength at any given temperature. However it had been discovered by Wilhelm Wien several years before Max Planck developed that more general equation, and describes the entire shift of the spectrum of black body radiation toward shorter wavelengths as temperature increases (Fig. 5.2).

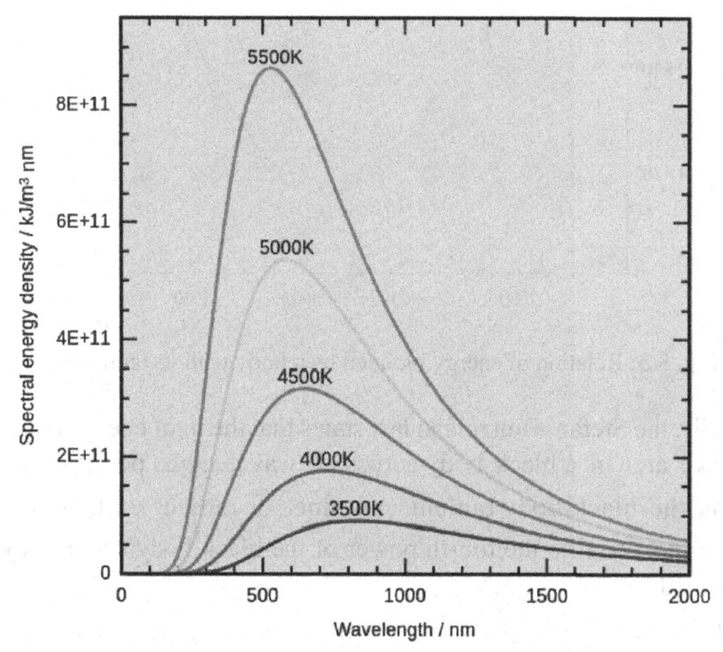

Fig. 5.2: Black body radiation as a function of wavelength for various absolute temperatures

Formally, Wien's displacement law states that the spectral radiance of black body radiation per unit wavelength, peaks at the wavelength λ_{max} given by:

$$\lambda_{max} = \frac{b}{T}$$

where T is the absolute temperature in kelvin. b is a constant of proportionality called Wien's displacement constant, equal to $2.8977729(17)\times10^{-3}$ m K, or more conveniently to obtain wavelength in micrometers, bH≈2900 μm·K.

Stefan–Boltzmann Law

The Stefan–Boltzmann law describes the power radiated from a black body in terms of its temperature (Fig. 5.3).

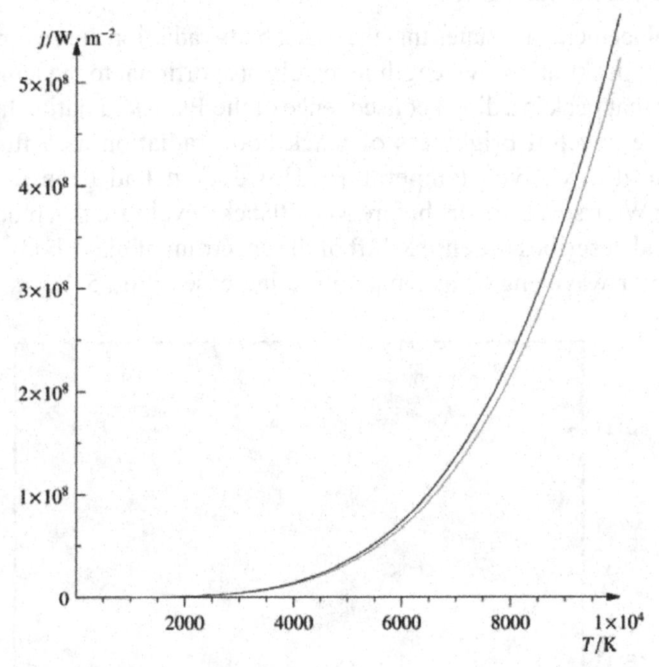

Fig. 5.3: Relation of energy radiated by a body with its temperature

Specifically, the Stefan–Boltzmann law states that the total energy radiated per unit surface area of a black body across all wavelengths per unit time (also known as the black-body radiant emittance or radiant exitance), j^*, is directly proportional to the fourth power of the black body's thermodynamic temperature T:

$$j^* = \sigma T^4$$

The constant of proportionality, σ, called the Stefan–Boltzmann constant derives from other known constants of nature. The value of the constant is

$$\sigma = \frac{2\pi^5 k^4}{15c^2h^3} = 5.67037 \times 10^{-8} Wm^{-2} K^{-4}$$

where k is the Boltzmann constant, h is Planck's constant, and c is the speed of light in a vacuum. Thus at 100 K the energy flux is 5.67 W m^{-2}, at 1000 K 56,700 W m^{-2}, etc.

Kirchhoff's law

Kirchhoff's law states that the ration of emissive power $\varepsilon(\lambda,T)$ of bodies to their absorptivity is independent of the nature of the radiating body. This ratio is equal to the emissive power of the black body $\varepsilon_0(\lambda,T)$ and depends on the radiation wavelength ë and on the absolute temperature T:

$$\frac{\varepsilon(\lambda,T)}{\alpha(\lambda,T)} = \varepsilon_0(\lambda,T)$$

Kirchhoff's radiation law is one of the fundamental laws of thermal radiation and does not apply to other types of radiation. According to Kirchhoff's radiation law a body that, at a given temperature, exhibits a stronger absorptivity must also exhibit a more intensive emission.

Solar PV technology

Photovoltaics, also called solar cells, are electronic devices that convert sunlight directly into electricity (Fig. 5.4). The modern form of the solar cell was invented in 1954 at Bell Telephone Laboratories. Today, PV is one of the fastest growing renewable energy technologies and it is expected that it will play a major role in the future global electricity generation. The photovoltaic effect is when two different (or differently doped) semiconducting materials (e.g. silicon, germanium), in close contact with each other generate an electrical current when exposed to sunlight. The sunlight provides the electrons with the energy needed to leave their bounds and across the junction between the two materials. This occurs more easily in one direction than in the other and gives one side of the junction a negative charge with respect to the other side (p-n junction), thus generating a voltage and a direct current (DC). PV cells work with direct and diffused light and generate electricity even during cloudy days, though with reduced production and conversion efficiency. Electricity production is roughly proportional to the solar irradiance, while efficiency is reduced only slowly as solar irradiance declines.

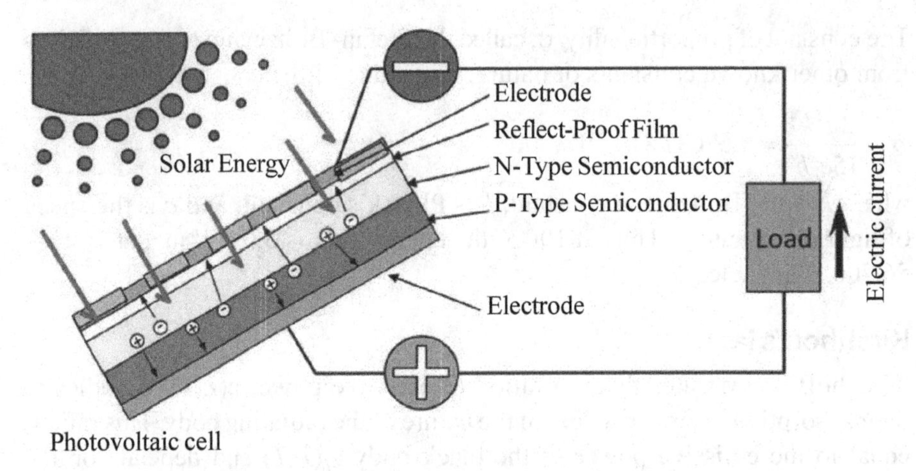

Fig. 5.4: The photovoltaic effect of electricity generation

A photovoltaic power generation system consists of multiple components like cells, mechanical and electrical connections and mountings and means of regulating and/or modifying the electrical output. These systems are rated in peak watts (W_p) which is an amount of electrical power that a system is expected to deliver when the sun is directly overhead on a clear day.

PV technology offers a number of significant benefits, including: (i) Solar power is a renewable resource that is available everywhere in the world; (ii) Solar PV technologies are small and highly modular and can be used virtually anywhere, unlike many other electricity generation technologies; (iii) Unlike conventional power plants using coal, nuclear, oil and gas; solar PV has no fuel costs and relatively low operation and maintenance (O&M) costs. PV can therefore offer a price hedge against volatile fossil fuel prices; (iv) PV, although variable, has a high coincidence with peak electricity demand driven by cooling in summer and year round in hot countries.

A PV system consists of PV cells that are grouped together to form a PV module, and the auxiliary components (i.e. balance of system - BOS), including the inverter, controls, etc. There are a wide range of PV cell technologies on the market today, using different types of materials, and an even larger number will be available in the future. PV cell technologies are usually classified into three generations, depending on the basic material used and the level of commercial maturity:

- First-generation PV systems (fully commercial) use the wafer-based crystalline silicon (c-Si) technology, either single crystalline (sc-Si) or multi-crystalline (mc-Si).

- Second-generation PV systems (early market deployment) are based on thin-film PV technologies and generally include three main families: 1) amorphous (a-Si) and micromorph silicon (a-Si/ìc-Si); 2) Cadmium-Telluride (CdTe); and 3) Copper-Indium-Selenide (CIS) and Copper-Indium-Gallium-Diselenide (CIGS).
- Third-generation PV systems include technologies, such as concentrating PV (CPV) and organic PV cells that are still under demonstration or have not yet been widely commercialised, as well as novel concepts under development.

First-generation PV technologies

Silicon is one of the most abundant elements in the earth's crust. It is a semiconductor material suitable for PV applications, with energy band gap of 1.1eV. Here the band gap indicates the energy needed to produce electron excitation and to activate the PV process. Crystalline silicon is the material most commonly used in the PV industry, and wafer-based c-Si PV cells and modules dominate the current market. The manufacturing process of wafer-based silicon PV modules comprises four steps: (i) Polysilicon production; (ii) Ingot/wafer production; (iii) Cell production; and (iv) Module assembly.

Crystalline silicon cells are classified into three main types depending on how the Si wafers are made. They are monocrystalline (Mono c-Si) sometimes also called single crystalline (sc-Si); polycrystalline (Poly c-Si), sometimes referred to as multi-crystalline (mc-Si); and EFG ribbon silicon and silicon sheet-defined film growth (EFG ribbon-sheet c-Si). Crystalline silicon technologies accounted for about 87% of global PV sales in 2010. The efficiency of crystalline silicon modules ranges from 14% to 19%.

Second generation PV technology: thin film solar cells

Thin-film solar cells could potentially provide lower cost electricity than c-Si wafer-based solar cells. Thin-film solar cells are comprised of successive thin layers, just 1 to 4 ìm thick, of solar cells deposited onto a large, inexpensive substrate such as glass, polymer, or metal. As a consequence, they require a lot less semiconductor material to manufacture in order to absorb the same amount of sunlight (up to 99% less material than crystalline solar cells). In addition, thin films can be packaged into flexible and lightweight structures, which can be easily integrated into building components (building-integrated PV, BIPV). The three primary types of thin-film solar cells that have been commercially developed are: Amorphous silicon (a-Si and a-Si/ìc-Si); Cadmium Telluride (Cd-Te); and Copper-Indium-Selenide (CIS).

Amorphous silicon solar cells, along with CdTe PV cells, are the most developed and widely known thin-film solar cells. Amorphous silicon can be deposited on cheap and very large substrates (up to 5.7 m^2 of glass) based on continuous deposition techniques, thus considerably reducing manufacturing costs. Currently, amorphous silicon PV module efficiencies are in the range 4% to 8%. The main disadvantage of amorphous silicon solar cells is that they suffer from a significant reduction in power output over time (15% to 35%), as the sun degrades their performance. A notable variant of amorphous silicon solar cells is the multi-junction thin-film silicon (a-Si/ìc-Si) which consists of a-Si cell with additional layers of a-Si and micro-crystalline silicon (μc-Si) applied onto the substrate. The advantage of the μc-Si layer is that it absorbs more light from the red and near infrared part of the light spectrum, thus increasing the efficiency by up to 10%. The thickness of the μc-Si layer is in the order of 3 μm and makes the cells thicker and more stable. Cadmium Telluride thin-film PV solar cells have lower production costs and higher cell efficiencies (up to 16.7%) than other thin-film technologies. This combination makes CdTe thin-films the most economical thin-film technology currently available. The two main raw materials are cadmium and tellurium. Cadmium is a by-product of zinc mining and tellurium is a byproduct of copper processing. Copper-Indium-Selenide (CIS) and Copper-Indium-Gallium-Diselenide (CIGS) PV cells offer the highest efficiencies of all thin-film PV technologies.

Third generation PV technology

Third-generation PV technologies are at the precommercial stage and vary from technologies under demonstration. There are four types of third-generation PV technologies:

- Concentrating PV (CPV);
- Dye-sensitized solar cells (DSSC);
- Organic solar cells; and
- Novel and emerging solar cell concepts.

Concentrating photovoltaic technology

Concentrating PV (CPV) systems utilise optical devices, such as lenses or mirrors, to concentrate direct solar radiation onto very small, highly efficient multi-junction solar cells made of a semiconductor material. The sunlight concentration factor ranges from 2 to 100 suns (low- to medium-concentration) up to 1000 suns (high concentration). To be effective, the lenses need to be permanently oriented towards the sun, using a single- or double-axis tracking system for low and high concentrations, respectively. Cooling systems (active or passive) are needed for some concentrating PV designs, while other novel approaches can get round this need. Low- to medium-concentration systems (up to 100 suns) can be combined with silicon solar cells, but higher temperatures will reduce their efficiency, while high concentration systems (beyond 500 suns) are usually associated with multi-junction solar cells made by semiconductor compounds from groups III and V of the periodic table (e.g. gallium arsenide), which offer the highest PV conversion efficiency. Multi-junction (either 'tandem' or 'triple' junction) solar cells consist of a stack of layered p–n junctions, each made from a distinct set of semiconductors, with different band gap and spectral absorption to absorb as much of the solar spectrum as possible. Most commonly employed materials are Ge (0.67 eV), GaAs or InGaAs (1.4 eV), and InGaP (1.85 eV). A triple-junction cell with band gaps of 0.74, 1.2 and 1.8 eV would reach a theoretical efficiency of 59%. Commercial CPV modules with silicon-based cells offer efficiency in the range of 20% to 25%. To maximise the electricity generation, CPV modules need to be permanently oriented towards the sun, using a single- or double-axis sun-tracking system. Multijunction solar cells, along with sun-tracking systems, result in expensive CPV modules in comparison with conventional PV.

Dye-sensitized solar cells

Dye-sensitized solar cells (DSSC) use photo-electrochemical solar cells, which are based on semiconductor structures formed between a photo-sensitised anode and an electrolyte. In a typical DSSC, the semiconductor nanocrystals serve as antennae that harvest the sunlight (photons) and the dye molecule is responsible for the charge separation (photocurrent). These cells are attractive because they use low-cost materials and are simple to manufacture. They release electrons from, for example, titanium dioxide covered by a light absorbing pigment. However, their performance can degrade over time with exposure to UV light and the use of a liquid electrolyte can be problematic when there is a risk of freezing.

Laboratory efficiencies of DSSC of around 12% have been achieved, however, commercial efficiencies are low - typically under 4% to 5%. The main reason why efficiencies of DSSC are low is because there are very few dyes that can absorb a broad spectral range. An interesting area of research is the use of nanocrystalline semiconductors that can allow DSSCs to have a broad spectral coverage. Thousands of organic dyes have been studied and tested in order to design, synthesise and assemble nanostructured materials that will allow higher power conversion efficiencies for DSSCs.

Organic solar cells

Organic solar cells are composed of organic or polymer materials (such as organic polymers or small organic molecules). They are inexpensive, but not very efficient. Organic PV module efficiencies are now in the range 4% to 5% for commercial systems and 6% to 8% in the laboratory. In addition to the low efficiency, a major challenge for organic solar cells is their instability over time. Organic cells can be applied to plastic sheets in a manner similar to the printing and coating industries, meaning that organic solar cells are lightweight and flexible, making them ideal for mobile applications and for fitting to a variety of uneven surfaces. This makes them particularly useful for portable applications, a first target market for this technology. Potential uses include battery chargers for mobile phones, laptops, radios, flashlights, toys and almost any hand-held device that uses a battery. The modules can be fixed almost anywhere to anything, or they can be incorporated into the housing of a device. They can also be rolled up or folded for storage when not in use. These properties will make organic PV modules attractive for building-integrated applications as it will expand the range of shapes and forms where PV systems can be applied. Another advantage is that the technology uses abundant, non-toxic materials and is based on a very scalable production process with high productivity.

Novel and emerging solar cell concepts

In addition to the above mentioned third-generation technologies, there are a number of novel solar cell technologies under development that rely on using quantum dots/wires, quantum wells, or super lattice technologies. These technologies are likely to be used in concentrating PV technologies where they could achieve very high efficiencies by overcoming the thermodynamic limitations of conventional (crystalline) cells.

Solar cell parameters

Typical I-V curve of a solar cell is presented in Fig. 5.5. The short-circuit current is the current through the solar cell when the voltage across the solar cell is zero (i.e., when the solar cell is short circuited). Usually written as I_{SC}, the short-circuit current is shown on the I-V curve at vertical axis. The open-circuit voltage, V_{OC}, is the maximum voltage available from a solar cell, and this occurs at zero current. The open-circuit voltage corresponds to the amount of forward bias on the solar cell due to the bias of the solar cell junction with the light-generated current. The open-circuit voltage is shown on the I-V curve below on horizontal axis. The short-circuit current and the open-circuit voltage are the maximum current and voltage respectively from a solar cell. However, at both of these operating points, the power from the solar cell is zero. The "fill factor", more commonly known by its abbreviation "FF", is a parameter which, in conjunction with V_{oc} and I_{sc}, determines the maximum power from a solar cell. The FF is defined as the ratio of the maximum power from the solar cell to the product of V_{oc} and I_{sc}. Graphically, the FF is a measure of the "squareness" of the solar cell and is also the area of the largest rectangle which will fit in the IV curve. The efficiency is the most commonly used parameter to compare the performance of one solar cell to another. Efficiency is defined as the ratio of energy output from the solar cell to input energy from the sun. In addition to reflecting the performance of the solar cell itself, the efficiency depends on the spectrum and intensity of the incident sunlight and the temperature of the solar cell. Therefore, conditions under which efficiency is measured must be carefully controlled in order to compare the performance of one device to another. The height of I-V curve also depends on the solar irradiation or insolation. Higher the insolation, the I-V curve shifts upwards and electric power produced by the solar cell increases.

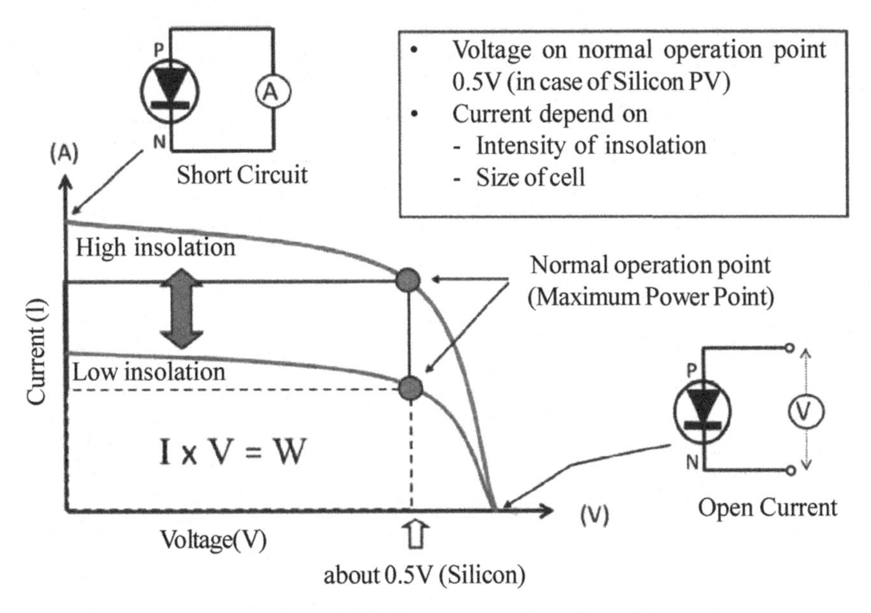

Fig. 5.5: Typical I-V curve of a solar cell

Summary

Four radiation laws e.g. Planck's law, Wien's displacement law, Steffan-Boltzaman law and Kirchhoff's law are described in the chapter. These laws govern the energy distribution of solar irradiation at different wavelengths in electromagnetic spectrum and its relation with the nature of radiating body. Further, the photovoltaic effect is discussed by which current is generated by a semiconductor when solar radiation falls on it. Different types of solar cells along with its brief characteristics are discussed thereafter. Finally different characteristics parameter of a solar cell is presented by illustrating the typical I-V curve of a solar cell.

6

Solar Cooker and Dryer Basic Design Criteria

N. M. Nahar

Chairman, Desert Science Institute, Jodhpur and Former Principal Scientist, ICAR-CAZRI, Jodhpur, Rajasthan, India

Introduction

Cooking accounts for a major share of energy consumption in developing countries. Fifty per cent of the total energy consumed in India is for cooking (Fritz 1981). Most of the cooking energy requirement is met by non-commercial fuels such as firewood (75%), agricultural waste and cow dung cake (25%) in rural areas. The fuel wood requirement is 0.4 tons per person per year in India. In rural areas firewood crisis is far graver than that caused by a rise in oil prices. Marginal villagers have to forage 8 to 10 hours a day in search of firewood as compared to 1 to 2 hours ten years ago. One third of India's fertilizer consumption can be met if cow dung is not burnt for cooking and instead is used as manure. The cutting of firewood causes deforestation that leads to desertification. Fortunately, India is blessed with abundant solar radiation (IMD, 1985). The arid parts of India receive maximum radiation *i.e.* 7600-8000 MJ m^{-2} per annum, followed by semi-arid parts, 7200-7600 MJm^{-2}, per annum and least on hilly areas where solar radiation is still appreciable i.e. 6000 MJm^{-2} per annum. Therefore, solar cookers seem to be a good substitute for cooking with firewood.

Principle of cooking

The different methods of cooking of food are boiling, frying, roasting, and baking. For boiling of rice, lentils etc. the temperature of food being cooked is about 100° C while for other methods, high temperatures are required. Heat is supplied at the bottom of the vessel for frying and boiling purposes in conventional cooking.

Roasting and baking is generally performed on open fire or in ovens, wherein food is surrounded by hot surfaces and heat is transferred to the food by radiation and convection.

The first step in cooking is to raise the temperature of the food to about 100°C that is the cooking temperature for most of the food in which water is present during cooking. If pressure cooker is used this temperature becomes 120°C. After food has attained cooking temperature, lesser heat is required to continue the cooking process. At this stage heat is supplied only to meet various heat losses taking place during cooking. The heat losses consist of evaporation loss from the food and convection and radiation losses from the surface of the cooking vessel. The relative proportion of these losses may very significantly depending upon the type of place (indoor or outdoor) of cooking.

Estimating an hourly convection loss (outdoors), at boiling water temperature of about 6.8 MJ m^{-2} of utensil and a surface area of 0.1 m^2 kg^{-1} of container contents, the energy inputs for 1 h of food boiling, if one fourth of the water present is vaporised, would be distributed roughly as follows (Lof 1961):

Heating food for boiling temperature	20%
Convection losses from the vessel	45%
Vaporization of water	35%

The heat losses can be reduced by insulating the sides of the cooking vessel and keeping the vessel covered with lid. Even though cooking temperature of most of the food is normally 100°C, the temperature of the heat source should be precisely higher to achieve satisfactory heat transfer rates. In the conventional cooking with direct fire, the heat transfer rate is very high, because of very high temperature of the heat source. Where electric or gas cooking is used, the normal burner supplies energy at the rate of approximately 1 kW and is capable of bringing 2 litre of water to boil in about 10 minutes. Therefore a solar unit should have to have an energy delivery rate of roughly 1 kW, to be compared with existing systems. The alternative would be accepting longer cooking periods and possibly cooking smaller amounts of food at one time. A solar cooker area of about 2 m^2 would be necessary (at 50% collection efficiency) to give comparable normal cooking rates.

Desirable features of a solar cooker

The solar cooker should meet following criteria:

(i) It must be possible for the user to obtain a cooking unit at a sufficiently low cost for him to realise financial saving by its use.

(ii) The cooker should cook all varieties of food effectively; it must therefore, provide energy at a sufficient rate and temperature to cook the desired quantities and type of food properly.

(iii) It must be sturdy enough to withstand rough handling and usage and to resist damage by natural hazards (such as wind) for the desired life time.

(iv) It must be capable of manufacturing with locally available materials and by local labour.

(v) The eyes and the hands of the housewife should be made safe from concentrated solar radiation.

(vi) A heat accumulator should make it possible to cook indoors and even after sunset.

(vii) The maintenance cost of the cooker should be low and tracking towards the sun should be as low as possible.

(viii) It should be dependable.

Keeping all this in mind several attempts have been made to develop an efficient, low cost and dependable solar cookers in various parts of the world. While all these requirements are necessary, there are some other important considerations which may determine the acceptability of solar cookers and these relates to the manner of use of solar cookers. The essential feature of a practical solar cooker have been described by Lawand (1973) as below:

(i) Adjustment of focus during the day should not be necessary. The person responsible for cooking meals should not be required to do anything other than placing the food and taking out of the solar cooker.

(ii) Food holding part of the cooker should be separate from the collectors.

(iii) The cooker should be rugged, durable and stable.

Since solar energy must be collected at temperatures at least 100°C to start the process of cooking and at higher temperatures to do baking and frying, it becomes necessary to have concentration of solar energy. Higher the concentration ratio provided, higher is the frequency of adjustment of the cooker. In order to cook all types of cooking, boiling, baking, and frying, it is necessary to have temperatures of cooking vessel in the range of 150°C and 200°C.

Brief history

The first solar furnace was fabricated by naturalist Georges Louis Leclerc Buffon (1707-1788). But Horace-de-Saussure (1740-1799) was first in the world to use the sun for cooking. His oven consisted of spaced glass blocks on top of blackened surface by an insulated box. The sunlight entered the box through the glass and was absorbed by the black surface. A temperature of 88°C was achieved. Augustin Mouchot, a French physicist, described a solar cooker in his book "La Chaleur Solaire" published in Paris, in 1869. He used a paraboloid concentrator to focus solar radiation on a cooking pot and noted that that the surface gave 'putrid fermentation' and 'unbearable smell' whereas the inside remains uncooked. He has also reported in the same book earlier work on solar cooking by English astronomer, Sir John Herschel, in South Africa, between 1834 and 1838. His oven was simply a black box that was buried in sand for insulation and was provided with double glass cover through which solar energy entered the box. A temperature of 116°C was recorded and vegetables and meat were cooked. Adams, an army officer, made India's first solar cooker in 1878 using plane mirrors arranged in an 8 sided pyramidical structure, 700 mm in diameter at the larger end. The mirrors reflected the rays on a cylindrical cooking utensil enclosed in a glass jar and he cooked vegetables as well as meat in it at Bombay (Adams 1878). Langley (1884) built a hot box and carried it on an expedition to Mount Whitney. In spite of the snow and frozen ground, the box could be used for cooking at high altitudes. Abbot (1939) built solar oven using cylindrical parabolic reflector to focus sunlight on to a blackened pipe enclosed in a glass tube. In this system the heat was absorbed first by the fluid in the tube and then conveyed to the cooking utensil placed in an insulated box. In this cooker, cylindrical parabolic reflector automatically tracked the sun by means of a clockworks mechanism. Since then different types of solar cookers have been developed all over the world. The solar cookers can be classified into three broad categories (i) Reflector/focusing type (ii) Heat transfer type and (iii) Hot box type. These are described below.

Reflector/Focusing Type

Parabolic reflector focuses the parallel rays of the sun to a small area that gives a very high concentration ratio and hence high temperature can be obtained. The reflecting surface may be of silvered glass or polished metal or aluminized mylar. If a very sharp focus is not required then a spherical reflector can be used instead of a parabolic reflector. A Fresnel reflector or a Fresnel lens can also be used for concentrating solar radiation. The main reflector/focussing type solar cooker are described below:

NPL paraboloid solar cooker

This cooker was developed at the National Physical Laboratory, New Delhi by Ghai (1953). It consists of a paraboloid reflector having 450 mm focal length. The reflector was spun from aluminium sheet to the desired shape and then anodised to protect from weather and to maintain reflectivity. In elevation paraboloid has a diameter of 1180 mm with 240 mm cut off horizontally across the top for a vertical height of 850 mm. The face area normal to the incident solar rays is 0.76 m^2 that is reduced to an effective area of 0.67 m^2 by the necessary attachments. The reflector is mounted on a stand that provides both azimuthal and elevation tracking. The support for the cooking utensil is a wire netting fixed to a steel ring that can be adjusted manually to provide a horizontal position for the vessel. The cooking utensil is a cylindrical brass vessel of 180 mm in diameter and 80 mm in height with a flanged ring at the top. The reflectance of the reflector is 0.75. It was observed that under clear calm days, 1.0 litre of water can be boiled within 25 minutes. Different types of food can be cooked and its output is equivalent to 450 W of electric hot plate. The cooker was manufactured on a large scale in India (Ghai et al. 1953) but a reflector type solar cooker did not become popular due to its inherent defects e.g. it required tracking towards sun every ten minutes, cooking could be done only in the middle of the day and only in direct sunlight, its performance was greatly affected by dust and wind, there was a danger of the cook being burned as it was necessary to stand very close to the cooker when cooking and the design was complicated.

Wisconsin solar cooker

This cooker was developed by Duffie et al. (1961) at the University of Wisconsin. It consists of a moulded plastic reflector that uses a draped-formed, high impact polystyrene shell of 1200 mm diameter and 1.5 mm thickness, stiffened at the rim with a ring of 12.5 mm diameter thin walled aluminium tubing. A reflective lining of aluminized mylar polyester film is applied to the shells with an adhesive, so that the clear film forms a protective covering over the specular surface. The cooker which has an effective area of about 1.0 m^2 delivers about 40 to 55 per cent of incident beam radiation to a cooking vessel of 180 mm in diameter; e.g. maximum delivery rate of 400-500 watts at an incident beam total energy of 1.0 kW on the unshaped reflector. The cooker is capable of boiling 900 ml of water in 13 minutes when 940 W average direct radiation is available.

Umbrella type solar cooker

This folding umbrella type solar cooker was developed by Lof and Fester (1961) in USA. The reflector is composed of a frame work, similar to an umbrella frame, covered with a metallized plastic film laminated to cloth. The reflector looks like an umbrella, when it is opened, with a highly reflecting lining. It is made with a light aluminium frame, has 16 ribs, and is covered with aluminized Mylerrrayon laminated cloth. Its diameter is 1150 mm and focal length 600 mm. The concentration ratio of the cooker is about 12. The net efficiency of the cooker was found to be 23 %. The cooker is equivalent to 400 W electric hot plate and different types of food can be cooked.

Multi-facet type solar cooker

A multi mirror durable solar cooker was designed by Tabor (1966). It consists of a twelve concave glass mirrors each having an area of 675 cm^2. These mirrors are arranged in three rows of five, four and three and are held in position by 12 circular rings made of iron rod and welded at their points of contact. The cooker is capable of boiling 1.84 litres of water within 22 minutes for a direct radiation of 1.0 kWm^{-2}. The cooker is equivalent to 550 W electric hot plate and different types of food can be cooked.

VITA Fresnel reflector type solar cooker

A Fresnel reflector type solar cooker was made by VITA (1962). The reflector is made of simple curved surfaces and is constructed of 3 mm masonite to which aluminized mylar has been cemented. The reflector is 1150 mm in diameter and has a focal length of 750 mm. It delivers 500 W to a focal spot of 150 mm diameter.

Sun basket

The sun basket solar cooker was developed by Von Oppen (1977). It is a very cheap basket type aluminium foil coated solar cooker that can cook food within 10 to 20 minutes. It is essentially a big basket, the inside of which is smoothened by paper mache. The paper mache and the basket are formed over a precise paraboloid made of plaster of Paris. The inside surface of the basket is covered with reflecting aluminium foil. Water and egg have been boiled as early as 7:00 h at Hyderabad. It can cook rice and dal within 20 minutes.

Dish solar cooker SK-14

Dish solar cooker (SK 14) has been developed by EG solar, an NGO of Germany, which is manufactured in India. The cooker is made of reflecting aluminium sheet. It can cook food for 10 to 15 people under the sun. Frying and chapatti-making is possible that cannot be performed in hot box sola cooker. The cooking time for various dishes is 20 to 30 minutes. It can save on fuel for up to 10 LPG cylinders annually on full use.

Solar steam cooking system

With technical assistance provided by M/s HTT GmbH of Germany and funding from GATE/ GTZ also of Germany, the first solar steam cooking system to cook for 1,000 people was developed and installed at the Brahma Kumaris' Ashram, Mt. Abu in Rajasthan. In 1997, this was the first solar steam cooking system based on scheffler solar concentrators in the world (Fig. 6.1). Its success has led to many more systems being installed in India.

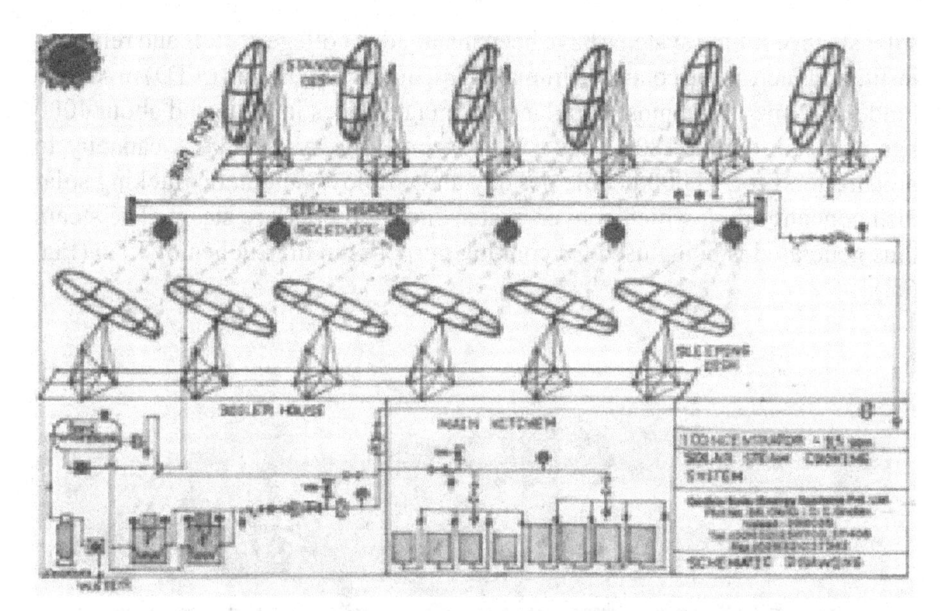

Fig. 6.1: Schematic of solar steam cooking system

Working principle of Solar Steam Cooking System

In the focus of each pair of scheffler concentrator (dishes), the sleeping dish and standing dish, are placed heat exchangers called receivers.

The solar rays falling onto the dish are reflected and concentrated on the receivers placed in its focus. Due to concentration the temperature achieved is very high (between 450-650°C) and thus the water in receivers comes to boiling and becomes steam.

Above the receiver is an insulated header pipe filled half with water. The cold water enters the receiver through inner pipe, gets heated due to the high temperature of the concentrated rays and the heated water goes up. The cold water again enters through inner pipe and the cycle continues till steam is generated. The steam gets stored in the upper half empty portion of the header pipe and pressure keeps on rising. The steam is than drawn / or sent to kitchen through insulated pipe line.

Spurred by the success of the above system, with training and jigs provided by Gadhia Solar, the Brahma Kumari's installed the world's largest solar steam cooking system at their Taleti Ashram in Abu Road, Rajasthan. This system installed in 1999, cooks upto 35,000 meals a day.

Around 25 systems with 5266 sq. m. of concentrated solar thermal area were also completed during the year 2015-16 making a total of 200 systems with 45,000 sq. m. of area installed so far in the country. In addition, a number of solar steam cooking systems have been installed at college hostels and religious institutions across the country. Tirumala Tirupathi Devasthanam (TTD) in Andhra Pradesh is one of the most popular pilgrimage places in India and about 4000 kgs of steam/day at 180°C and 10 kg/sq cm. The system has a capacity to prepare food for 15,000 people per day and employs automatic tracking solar dish concentrators, which convert water into high pressure steam. The steam thus generated is being used for cooking purposes in the kitchen of TTD (Fig. 6.2).

Fig. 6.2: Tirumala tirupati devasthanam steam cooking system

Solar Bowl for Cooking

The system which has been developed and installed at the Centre for Scientific Research (CSR), Auroville (Pondicherry) consists of a non-tracking solar bowl concentrator of 15 m diameter integrated with the terrace of CSR kitchen and a cylindrical automatic tracking receiver pivoted at its focal point from one end. The system uses thermic fluid to the energy collected by the receiver, which is stored in a heat storage tank. A heat exchanger fitted in this tank then generates steam, which is used for cooking food using double jacketed cooking pots. The oil can be heated up to 260°C, which is sufficient to generate steam for cooking food in the kitchen. Around 600 kg of stem per day could be generated from this bowl that is sufficient to cook two meals for about 1000 people. The system is also hybridised with a diesel-fired boiler capable of producing 200 kg of steam per hour so to ensure that the meals are ready in time despite unsuitable weather conditions.

Heat Transfer Type

In the heat transfer type solar cooker, the collector is kept outside and the cooking chamber is kept inside the kitchen of the house (Abot 1939; Alward 1972; Garg and Thanvi 1977). But this type of solar cooker also did not become popular because of its high cost and only limited cooking can be performed.

Hot Box Type

The third type of cooker is known as hot box in which most of the defects of above two types of cookers have been rectified (Ghosh 1956; Telkes 1959; Garg 1976; Nahar 1990; Grupp et al 1991; Nahar et al. 1994). Different types of solar cookers have been tested and the solar oven (Garg et al. 1978; Malhotra et al. 1983; Nahar 1986; Olwi and Khalifa 1988) has been found best. Different hot box solar cookers are described below:

Solar Oven

The solar oven consists of a reflector assembly cooking chamber and an angle iron stand. The cooking chamber is double walled cylindrical vessel. Cylinders are made from 22-gauge aluminium sheet. Space between them (100 mm) is filled by glass wool insulation. Two clear window glasses are fixed over it. The inner cylinder is painted black by black board paint. One door is provided for loading and unloading of cooker for cooking. It is fixed over an angle iron stand, which is having castor wheel for azimuthal tracking and a slotted kamani for elevation tracking. Reflectors are trapezoidal in shape and also made from 22-gauge aluminium sheet, are fixed over it. It consists of four rectangular and four triangular mirrors. One rubber gasket has been fixed to the boundary of the door to prevent the leakage of hot air and to increase the pressure inside the chamber enabling a reduction of cooking time. To facilitate azimuthal tracking, a stand of mild steel angle having four castor wheels has been made. A slotted kamani has been fixed for following the altitude position of the sun. The cooking utensils are kept on a cradle like platform so that vessels always remain in a horizontal position, irrespective of the oven's inclination. Performance of cooker was carried out by measuring stagnation plate temperature, time taken for known quantity of water for reaching boiling point and cooking trials. Cooking of vegetable, rice, roasting of potato, baking of bati is possible from 8.00 h to 17.00 h in winter while 7.00 h to 18:00 h in summer. Actual installation of the solar oven is shown in Fig. 6.3.

Fig. 6.3: Solar Oven

Hot box solar cooker

Though the performance of the solar oven is very good but it also requires tracking towards sun every 30 minutes, it is too bulky and is costly. Therefore, the hot box solar cooker with a single reflector (Parikh and Parikh 1978) is being promoted at subsidised cost by the Ministry of Non-conventional Energy Sources, Government of India and the state nodal agencies in India since 1981-82 and 1.33 million humber of solar cookers were sold up to December 2016 (MNRE, 2016). This is a double walled chamber outer tray is made from 20 gauge mild steel and inner tray is from 22 gauge aluminium sheet glass wool insulation. The inner tray is painted by black board paint. Two clear window glasses (3 mm thick) fixed in a wooden frame are hinged over it which can be lifted for loading and unloading of cooker for cooking (Fig. 6.4). One mirror booster with slotted kamani on top, which acts as a lid as well, has been provided. Four castor wheels are fixed at the bottom for easy movement. The four cooking utensils are provided in the cooking chamber for cooking four dishes simultaneously. The operation of the cooker is very simple. The products to be cooked are placed in the cooking utensils with right amount of water so that after cooking whole water is absorbed. The lid is closed and utensils are kept inside the cooking chamber. The booster mirror is adjusted in such a way that all reflected solar radiation by it falls on plain glass. The tracking of cooker towards sun should be done every hour. All types of boiling, bakery and roasting operation can be performed and it takes about 2 to 3 hrs in cooking one kg of food in four utensils. Soft food takes less time hard food takes more time.

Fig. 6.4: Hot box solar cooker

Improved hot box solar cooker with tilted absorber

The performance of hot box solar cooker is very good during summer but it is very poor during winter in northern parts of India and difficult to cook two meals per day during winter because its glass window and absorbing surface are horizontal, which receive very much less radiation as compared to optimally inclined surface. Optimally inclined surface receives 43.8% and 22.8% more radiation as compared to horizontal surface during winter (October to March) and per year respectively. Considering this, a novel solar

Fig. 6.5: Improved Hot Box Solar Cooker with Tilted Absorber

cooker with tilted absorber (TA) has been designed, fabricated and tested (Fig. 6.5). It has been found that the performance of solar cooker (TA) is better than hot box solar cooker and comparable with solar oven, and simultaneously no tracking is required as compared to 30 minute tracking for solar oven and 60 minute tracking for the hot box solar cooker. The overall efficiency of the solar cooker (TA) has been found to be 24.6%. Cooking trials have also been carried out at different times with different materials and the time taken to cook various dishes is between 75 and 120 minute for the solar oven, 90-180 minute for the hot box and 90-150 minute for the solar cooker (TA). If the cookers are partially loaded, then cooking time is less. Cooking time is less around noon time and, while it is more in the morning and the evening.

Hot box solar cooker with transparent insulation material (TIM)

The performance of hot box solar cooker is very poor during winter in northern parts of India and difficult to cook two meals per day during winter because more heat loss due to low ambient temperatures. It has been found that the efficiency of the solar devices can be increased considerably for temperature applications between 80°C to 140°C *i.e.* for solar cooking, process heat and for refrigeration applications by using transparent insulation material (TIM) in between two glazing or between absorber and glazing. The use of TIM reduces convective heat losses from glass window. Considering this, a hot box solar cooker with 40 mm thick TIM encapsulated between two glazing has been designed, developed and tested (Fig. 6.6). The efficiency of the cooker with TIM is 30.4% as compared to 15.7 % without TIM. Cooking trials have also been carried out at different times with different materials. One kg of dry food can be cooked in 2 hour and 2.5 hour in hot box solar cooker with and without TIM respectively.

Fig. 6.6: Hot box solar cooker with transparent insulation material (TIM)

Double reflector hot box solar cooker with a transparent insulation material (TIM)

The popularity of hot box solar cooker promoted by MNRE, New Delhi and state nodal agencies is declining due to its defects: it requires tracking towards the sun every 60 minutes, therefore, its operation becomes cumbersome and the performance of the hot box solar cooker is very poor during winter when solar radiation and ambient temperatures are very low. Considering this, double reflector hot box solar cooker with a transparent insulation material (TIM) has been designed, fabricated, tested and the performance has been compared with a single reflector hot box solar cooker without TIM (Fig. 6.7). Both defects of the hot box solar cooker have been removed by providing one more reflector and convective heat losses have been suppressed by using transparent insulation material (TIM). The efficiencies were 30.5% and 24.5% for cookers with and without a TIM respectively during winter season at Jodhpur. The performance studies on the double reflector hot box solar cooker with TIM suggests that the cooker can be used throughout the year.

Fig. 6.7: Double reflector hot box solar cooker with a transparent insulation material

Non-tracking solar cooker

The performance of the hot box solar cooker is good but it requires tracking towards sun every 60 minutes, therefore, its operation also becomes cumbersome. To eliminate tracking completely, a novel non-tracking solar cooker has been designed, developed and tested. The cooker is based on hot box principle having a single reflector. The cooker was designed in such a way that the length to width ratio of the cooker has been fixed as 3:1 so that maximum amount of radiation falls on the glass window any time during the day. It helped in eliminating the need for azimuthal tracking of the cooker, which is very essential for a simple hot box solar cooker, towards the sun every hour because the length to width ratio of reflector is 3:1. The outer box is made of galvanised steel sheet (22 SWG) and inner of aluminium (22 SWG). The space between the outer box and inner box was filled with glass wool insulation. The inner tray is painted black using black board paint. Two clear window glass planes of 4 mm thickness have been fixed over it with an openable wooden frame. A 4 mm thick plain mirror reflector is fixed over it. The tilt of the reflector can be varied from 0° to 120° depending upon the season and its tilt is fixed once in a fortnight. The reflector was folded on the cooker while the device is not in use. The aperture area of the solar cooker is 0.25 m^2. Three cooking utensils of aluminium/ stainless steel boxes with lid can be kept inside it for cooking four dishes simultaneously. The cooker is fixed on an angle iron stand. Actual

Fig. 6.8: Non-Tracking Solar Cooker

installation of the non-tracking solar cooker is shown in Fig. 6.8. The performance of the non-tracking solar cooker is comparable with the hot box, though it is kept fixed while the hot box is tracked towards sun every hour. It has been made possible because the width to length ratio is 3 for the non-tracking solar cooker, while it is 1 for the hot box solar cooker. Cooking trials have also been conducted and rice, lentils, kidney beans, cauliflower, backing of *bati* (local preparation made of wheat flour) etc. have been cooked successfully. It takes about 2h for soft food and 3h for hard food. The cooker is capable of cooking 1.0 kg of food at a time. The efficiency of the non-tracking solar cooker has been found to be 29.5 %.

Hot box storage solar cooker

Solar cookers are used only during the day. To overcome this problem, a hot box solar cooker with used engine oil as a storage material has been designed, fabricated, and tested so that cooking can be performed even in the late evening. Performance and testing of a storage solar cooker has been carried out by measuring stagnation temperatures and conducting cooking trials. The maximum stagnation temperature inside cooking chambers of hot box solar cooker with storage material was 136°C same as of hot box solar cooker without storage during the day time but it was 23°C more in storage solar cooker from 17:00 h to 24:00 h. The cooking trials were also conducted. The rice and green gram washed split were kept at 17:30 h and these were cooked perfectly by 20:00 h in hot box solar cooker with storage while these were not cooked in hot box solar cooker without storage.

Community solar cooker

A community solar cooker capable of cooking for about 80 persons has been designed, fabricated and tested at CAZRI, Jodhpur. The cooker is suitable for hostels, temples, canteens, restaurants, etc. The cooker is based on the hot box principle having a single reflector. The cooker has been designed in such a way that the width to length ratio for the reflector and the glass window is about 4, so maximum radiation falls on the glass window. This has helped in eliminating the azimuthal tracking, which is required in the simple hot box solar cooker, towards the sun every hour because the width to length ratio of the reflector is 1. This cooker is always kept fixed, facing the equator. This device consists of a double walled hot box. The outer tray is made of mild steel and the inner of aluminium. The space between them is filled with glass wool insulation. The inner tray is painted by blackboard paint. Two clear window glass panes of 4 mm thickness have been fixed over it with a wooden frame. Three doors have been provided in the rear side for loading and unloading the cooker. The doors have been made leak proof by rubber gaskets. A 4-mm thick plain mirror reflector is fixed over it and arrangements have been made so that it can be tilted to 120° from the glass window. Therefore, it is effective in summer as well as in winter when the altitude of the sun is very low. The absorber area of the cooker is 3.12 m^2. Specially designed cooking utensils are rectangular in shape, having dimension 560×540×75 mm^3. These are made from aluminium sheet. Twelve such utensils can be kept inside the cooker. The cooker was tested extensively (Fig. 6.9).

Fig. 6.9: Community solar cooker

The stagnation air temperature inside the cooking chamber has been measured and compared with the hot box solar cooker. The maximum stagnation temperature inside cooking chamber during summer is 146°C and 136 °C in winter. The efficiency of the cooker is 28.4 %. The cooker can be used for boiling, roasting and baking. The cooker is capable of cooking 16 kg of food at a time. The performance of the community solar cooker is comparable with the hot box, though it is kept fixed while the hot box is tracked towards the sun every hour. It has been made possible because the width to length ratio is 4 for the community solar cooker, while it is 1 for the hot box solar cooker.

The cooker can be used twice a day for about 254 days and once a day for about 67 days in a year at Jodhpur. The energy for cooking per person is about 900 kg of fuel equivalent per meal. The community solar cooker is capable of cooking for about 80 persons, and it will save 50 % of cooking fuel per meal. Therefore, it will save 36 MJ of energy per meal and 20,700 MJ of fuel equivalent per year. The cooker can also be used in cottage industries for preparation of rose syrup, gulkand, processing of *anwala, ber* etc. for jam and jelly making.

Animal feed solar cooker

During the survey of rural arid areas of Rajasthan, it was found that huge amount of firewood, cow dung cake and agriculture waste is burnt for boiling of animal feed. The feed is generally given to the animals in the evening. The solar cookers available are suitable for cooking food twice a day, therefore, their cost is high while animal feed is to be boiled only once a day. Therefore, it was felt that a very low cost suitable solar cooker should be designed for boiling of animal feed.

Large size animal feed solar cooker made of clay

The performance of small size solar cooker for animal feed made of clay is very good but its capacity is only 2 kg per day while in western Rajasthan farmers have 4 to 5 cattle. Therefore, a large size animal feed cooker has been designed, fabricated, and tested for boiling of 10 kg of animal feed per day (Fig. 6.10). The cooker employs locally available materials of no cost e.g. clay, pearl millet husk and cow dung. The commercial material for its fabrication are plain glass, mild steel angle and sheet, wood and aluminium sheet cooking utensils. The frame body of the cooker may be Fabricated by an unskilled labour. The cooker is capable of boiling 10 kg of animal feed, sufficient for five cattle per day. The efficiency of the cooker is 21.8%. The cooker saves 6750 MJ of energy per year.

Fig. 6.10: Solar cooker for animal feed at the village Osian, Jodhpur

Large size animal feed solar cooker made of vermiculite-cement

The performance of animal feed made of clay is very good. By observing its success it was felt that more durable solar cooker should be designed by using cement and vermiculite. The performance and testing of this cooker has been compared with solar cooker made of clay etc. The cooker was tested by measuring stagnation plate temperature and it was observed as high as 120°C as compared to 110°C observed in solar cooker made of clay. Different types of animal feed e.g. crushed barley (local name "Jau Ghat"); cluster bean split, cluster bean powder, gram powder etc. have been tried. Three aluminium cooking utensils each having 3.0 kg of crushed barley/cluster bean with 8.0 litre water were put inside cooking chamber at 9:00 h and it was cooked perfectly by 14:00 h. These animal feed are commonly used in the Thar desert of Rajasthan. The animal feed is generally given to the animals between 16:00 h to 20:00 h. The efficiency of solar cooker for animal feed has been found to be 26.4%. Actual installation of the large size animal feed solar cooker made of vermiculite cement is shown in Fig. 6.11.

Fig. 6.11: Animal feed solar cookers

Solar tea boiler

A solar tea boiler has been designed, developed and tested (Fig. 6.12). The device can be used to boil 125-150 cups of tea from 10:00 h to 17:00 h. The device was tested for boiling of water and milk. The boiled water was collected every ten minutes from the device and observations were recorded from 10 Am onwards on clear as well as on cloudy days. The efficiency of the device was found to be 34.2%. On an average device can be used to boil 16.5 litres of water and rise in water temperature was observed to be 65°C. The operating temperature being 95°C or more. It was observed that 300 clear sunny days are available in most parts of India accordingly calculation of fuel saving has been made and it was found that device saves 677 kg of firewood or 492 kWh of electricity or 169 kg of coal or 78 litres of kerosene.

Fig. 6.12: Solar tea boiler

Thermal performance and testing of solar cooker

A procedure for testing the solar cookers was developed based on existing international testing standards. These include three major testing standards for solar cookers that are commonly employed in different parts of the world; (i) American Society of Agricultural Engineers Standard (ASAE 2003), (ii) Bureau of Indian Standards Testing Method (BIS 1992 and 2000), and (iii) European Committee on Solar Cooking Research Testing Standard and others (ECSCR

1994). Based on the existing international testing standards three tests were performed on the high insulation box type solar cooker, these are: first figure of merit, F_1 and second figure of merit F_2. The first figure of merit (F_1) is determined by conducting the no-load test, second figure of merit (F_2) was determined by load test in which known amount of water is sensibly heated in solar cooker and cooking power estimation.

First figure of merit (F_1) without water load (stagnation test): The first figure of merit (F_1) is defined as the ratio of optical efficiency, (η_0), and the overall heat loss coefficient, (U_L). A quasi-steady state (stagnation test condition) is achieved when the stagnation temperature is attained. High optical efficiency and low heat loss are desirable for efficient cooker performance. Thus the ratio η_o/UL which is a unique cooker parameter can serve as a performance criterion. In stagnation test initially temperature of bare plate increases and after some time it gets stagnant. Higher values of F_1 would indicate better cooker performance (Mullick et al. 1987):

$$F_1 = \frac{\eta_0}{U_L} = \frac{(T_{ps} - T_a)}{G_s}$$

Where F_1 is first figure of merit, η_o is optical efficiency (%), UL is overall heat loss coefficient of the cooker (W/m^2 °C), T_{ps} is maximum plate surface temperature (°C), $\overline{Ta}$ is ambient temperature (°C), and $\overline{G}_s$ is global solar radiation on a horizontal surface (W/m^2).

Second Figure of Merit (F_2) with water load (Sensible heat test): The second figure of merit, F_2, of box type solar cooker is evaluated under full-load condition (water load), without reflector and can be defined as the product of the heat exchange efficiency factor (F`), optical efficiency ($\eta_o = \alpha\tau$) and heat capacity ratio (C_R). It can be expressed as (Mullick et al. 1996):

$$F_2 = F'\eta_0 C_R = \frac{F_1(MC)w}{A(t_2 - t_1)} ln \left[\frac{1 - \frac{1}{F_1}\left(\frac{T_{w1} - \overline{T}_a}{\overline{G}_s}\right)}{1 - \frac{1}{F_1}\left(\frac{T_{w2} - \overline{T}_a}{\overline{G}_s}\right)} \right]$$

where F_1 is first figure of merit (°C m^2/W), $(MC)_w$ is product of the mass of water and its specific heat capacity (J/ °C), A is aperture area of the solar cooker (m^2), t_1 is initial time (s), t_2 is final time (s), T_{w1} is initial water temperature (°C), T_{w2} is final water temperature (°C), $\overline{G}_s$ is average global solar radiation (W/m^2), $\overline{T}_a$ and is average ambient temperature (°C).

Efficiency of the solar cooker (h)

Performance of solar cooker has been carried out extensively by measuring stagnation plate temperature and rise in water temperature in cooking utensils in known interval of time. The stagnation plate temperature was measured by putting four numbers of thermocouples on the plate and on air inside cooking chamber and temperature of each was measured by the portable digital thermometer with suitable sensor (accuracy 0.1°C) and average of initial and final were taken. The initial temperature of cold water was measured and when it reached near the boiling point temperature of water, the final temperature of hot water was measured and time interval was also measured. The efficiency of the cooker has been found by the following relations proposed by Nahar (2001 and 2009):

$$\eta = \frac{(MC_w + M_1C_u)(T_{w2} - T_{w1})}{CA\int_0^t G dt}$$

Where A = Absorber area (m^2); C = Concentration ratio; C_u = Specific heat of cooking utensil (J/kg/°C); C_w = Specific heat of water (J/kg/°C); G = Solar radiation (W/m^2); M = Mass of water in cooking utensils (kg); M_1 = Mass of cooking utensils (kg); T_{w1} = Initial temperature of water (°C); T_{w2} = Final temperature of water (°C); t = Time interval (s) and η = Efficiency of solar cooker (%).

Solar Dryer

Drying or dehydration of material means removal of moisture from the interior of the material to the surface and then to remove this moisture from the surface of the drying material. The product is directly exposed to the sun in the open air in natural sun drying. The necessary heat for removal of moisture is supplied by the sun while in the convection type dryers, a stream of preheated air form solar energy supplemented by auxiliary energy is allowed to pass through the product which supplies the necessary heat for moisture removal from inside to outside and also carries the moisture.

In many rural areas of India, the farmers grow fruit and vegetables. These perishable commodities have to be sold in the market immediately after harvesting. When the production is high, the farmers have to sell the material at very low price, there by incurring great loss. This loss can be minimised by dehydrating fruits and vegetables. The dried products can be stored for longer time in less volume. In off seasons the farmer can sell the dried products at higher price. The traditional methods for drying the agricultural produce is to dehydrate the material under direct sunshine. This method of drying is a slow process and usual problems like dust contamination, insect infestation and spoilage due to unexpected rain. These problems can be solved by using either oil-fired or gas fired or electrically operated dryers. However, in many rural locations in India, the electricity is either not available or too expensive for drying purpose. Thus in such areas the drying systems based on the electrical heating are inappropriate. Alternatively, fossil powered dryer can be used but it poses such financial barriers due to large initial and running cost that these are beyond the reach of small and marginal farmers. In the present energy crisis, it is desirable to apply a little solar technology for dehydration of fruits and vegetables, so that gas, oil and electricity can be saved. India is blessed with abundant solar energy, which can be used for dehydrating fruits & vegetables through solar dryer. Keeping this in view, solar dryers both direct type viz. simple solar cabinet dryer, improved dryer with chimney, dryer for maximum energy capture, multirack tilted type dryer, low cost dryer and forced circulation type dryer have been designed, developed and tested at Central Arid Zone Research Institute (CAZRI), Jodhpur (Thanvi and Pande 1989). There are different types of solar dryers, which are described below.

Natural convection or direct type solar dryers

Solar cabinet dryer

The solar cabinet dryer consists of a wooden or metallic box where length to width ratio is kept 3, insulated at the base and sides and covered with a single glazing (Fig. 6.13). The inside surface of the dryer is painted black by black board paint. The product to be dried is kept on wire mesh trays. These trays can be kept inside dryer by an openable door provided in the rear side of the dryer. Ventilation holes are made in the bottom through which fresh outside air is sucked. A chimney with a regulating valve is provided on the top of the rear side for escaping the moisture from the product to the environment. Different types of agriculture produce can be dehydrated; it takes about 5 to 7 days for different products.

Fig. 6.13: Solar cabinet dryer

Commercial solar dryer with inclined surface

In many rural areas of India, the farmers grow fruit and vegetables. These perishable commodities have to be sold in the market immediately after harvesting. When the production is high, the farmers have to sell the material at very low price, there by incurring great loss. This loss can be minimised by dehydrating fruits and vegetables. The dried products can be stored for longer time in less volume. In off-seasons, the farmer can sell the dried products at higher price. The traditional methods for drying the agricultural produce are to dehydrate the material under direct sunshine. This method of drying is a slow process and usual problems like dust contamination, insect infestation and spoilage due to unexpected rain. These problems can be solved by using either oil-fired or gas fired or electrically operated dryers. However, in many rural locations in India, the electricity is either not available or too expensive for drying purpose. Thus in such areas the drying systems based on the electrical heating are inappropriate. Alternatively, fossil powered dryer can be used but it poses such financial barriers due to large initial and running cost that these are beyond the reach of small and marginal farmers. In the present energy crisis, it is desirable to apply a little solar technology for dehydration of fruits and vegetables, so that gas, oil and electricity can be saved. India is blessed with abundant solar energy, which can be used for dehydrating fruits & vegetables through solar dryer. The details of the commercial type solar dryer are described below.

A large size solar dryer (capacity 100 kg, glass area 10 m^2) which can be commercially used for drying fruits and vegetables was developed. The salient features of this dryer are: (i) it can capture the maximum solar energy throughout the year by keeping the system at optimum tilt during different seasons. (ii) It can protect the drying material from rain, flies and squirrel. (iii) Stainless steel wire mesh is used for fabrication of drying trays. (iv) Partitions are provided in the drying trays so that the material can be stacked even on inclined plane. (v) A low cost material viz. bajra (*pearl millet*) stem is used as insulation. (vi) The dryers can be connected in series and hence its capacity can be enhanced as per requirement and (vii) it can be dismantled easily so that its transportation is easy from one place to another. Actual installation of the inclined solar dryer with spinach is shown in Fig. 6.14.

Drying trials for dehydrating vegetables viz. mint, spinach, okra, tomato, ginger, red and green chillies, carrot, coriander leaves, fenugreek, peas, cabbage, onion, sweet potato, bitter gourd, radish, sugar beet, cauliflower, bathua and fruits, viz. ber (*Indian jujube*), sapodilla, grapes, pomegranate, etc. were conducted successfully. The detailed results of some drying trials are given in Table 6.1.

Fig. 6.14: Commercial solar dryer with inclined surface

The results indicated that the leafy vegetables can be dehydrated within 2 to 3 days at the loading rate of 4 to 5 kg/m^2, whereas other vegetables can be dried within 3-4 days at loading rate of 8 to 10 kg/m^2. Thus in general, it can be concluded that in commercial solar (glass area 10 m^2) about 100 kg of vegetables can be dried in 4 days. The green colour of solar dried products remained as such even after drying. These solar dried vegetables should be soaked in hot water before cooking. The spinach powder can be used for making 'Palak paneer'. The coriander and tomato powder can be mixed with ingredients to prepare instant soup/sauce/chutney by adding water. The solar dried grated carrot can be used for preparing pudding 'Gajar ka Halwa'.

Table 6.1: Drying trials in solar dryer

Item	Loading Rate (kg m^{-2})	Drying time (days)
Spinach	4.5	2
Coriander	4.0	2
Mint	3.0	1.5
Okra	10.0	3
Green chilli	10.0	3.5
Tomato	5.0	2
Sweet potato	11.0	3.5
Fenugreek leaves	5.2	1.5
Cabbage	7.3	2.6
Bitter gourd	4.5	2.1
Sugar beet	8.0	2.2
Carrot (Gratin)	8.0	2.2
Ginger (Gratin)	10.0	2.4
Peas	6.7	2.5
Ber	15.0	3.5

Thermal Efficiency (η)

The efficiency of utilization of solar energy in solar dryer (ratio of heat used in evaporation of moisture from the fruit to the incident total solar radiation on horizontal plane) was worked out using the following relation (Leon et al. 2002; Poonia et al. 2018):

$$\eta = \frac{ML}{A\int_0^\theta H_T d\theta}$$

Where,

A = Absorber area (m^2)

H_T = Solar radiation on horizontal plane ($J\ m^{-2}\ h^{-1}$)

L = Latent heat of vaporisation ($J\ kg^{-1}$)

M = Mass of moisture evaporated from the product (kg)

θ = Period of test (h)

η = Efficiency of the solar dryer

Advantages of solar dryers

- Solar dryer can save fuel and electricity as required in case of mechanical drying method.
- Drying time in solar dryer is reduced in comparison to open drying method.
- Fruits and vegetables dried in solar dryer are better in quality and hygienic than dried in open.
- The limited space available in houses in large cities can be effectively used for dehydrating fruits and vegetables using domestic solar dryer.
- Materials required for fabrication of solar dryer are locally available.
- Use of solar dryer involves no fire risks.
- The trade of dried vegetables can be linked with national and international trades.

In the mixed mode type dryers the solar air heater without any fan along with the drying bin is used. The flow of air is maintained by natural convection. This type of dryer known as rice dryer was developed by AIT, Bangkok, Thailand. It consists of a simple air heater, drying chamber and a tall chimney used to increase the convection effect. The air heater is made of a frame of bamboo poles and covered with a 0.15 mm thick PVC sheet. The ground is covered with burnt rice husk, which absorbs the solar radiation and heats the air in contact. The hot air rises to the drying chamber, which either consists of a transparent PVC sheet on bamboo frame absorbing directly the solar radiation. The drying material is kept on a nylon net tray in thin layer through which hot air heated from air heaters enters its bottom and goes up into the chimney. The chimney is made from bamboo frame and is cylindrical in shape and covered with black PVC to keep inside air warm. The height of the chimney and the hot air inside creates a pressure difference between its top and bottom thereby creating forced movement of air through a pressure its top and bottom. It creates forced movement of air through the product bed to the chimney.

Solar timber kiln

The timber kiln consists of a large green house generally rectangular with all the four walls and roof made of single or double-glazing either of plastic or glass. The concrete floor is bit raised for stacking timber in such a way that there is sufficient space for circulation of air. An electrically operated fan is used for circulating the air inside the chamber through the stack. The inside of the kiln is painted black by black board paint. The timber is dried in the kiln efficiently.

Forced circulation type solar dryer

In these dryers, blower is used for the circulation of air which is either operated electrically or mechanically. These dryers are more efficient, faster and can be used for drying large quantity of agricultural products. There are two types of forced circulation dryers (1) Direct mode (2) Indirect mode. The direct mode is similar to the indirect type natural convection dryer except a blower makes flow of air. These are not efficient. But indirect type or forced circulation is very efficient. In these dryers airflow is obtained by blower, air gets heated in the air heaters; hot air is passed through the product kept inside a bin where air takes moisture from the product and escape out.

References

Abot, C.G . 1939. Smithsonian Misc. Cells

Adams W. 1878. 'Cooking by solar heat', Scientific American, 38: 376

Alward, R. 1972. 'Solar steam cooker Do it yourself leaflet L-2', Brace Research Institute, Quebec, Canada.

ASAE. 2003. ASAE S580: Testing and Reporting of Solar Cooker Performance.

Bureau of Indian Standards. 1992. BIS standards on solar cooker IS 13429, Part I, II and III, Manak Bhavan, New Delhi, India.

Bureau of Indian Standards. 2000. BIS standards on solar – box type cooker - IS 13429, Part I, II and III, Manak Bhavan, New Delhi, India.

Duffie, J.A., Lof, G. O. G. And Beck, B. 1978. 'Laboratories and field studies of plastic reflector solar cookers', Proceedings of UN conference on New sources of Energy, Rome, Paper S/87, 5: 339-346

ECSCR. 1994. Solar Cooker Test Procedure: Version 3.

Fritz, M. 198. Future Energy consumption, Pergamon Press, New York

Garg H.P. and Thanvi K.P. 1977. Studies on solar steam cooker. Indian Farming 27(1): 23-24.

Garg, H.P. 1976. A solar oven for cooking, Indian Farming 27: 7-9

Garg, H.P., Mann, H.S. and Thanvi K.P. 'Performance evaluation of five solar cookers', Proc. ISES Congress, New Delhi (Eds. F de Winter and M. Cox) Pergamon Press New York SUN 2:1491-1496, 1978.

Ghai, M.L. 1953. Design of reflector type direct solar cooker.Journal of Scientific and Industrial Research 12 A: 165-175.

Ghai, M.L., Pandhar, B.S. and Harikishan, D. 1953. Manufacture of reflector type direct solar cooker. Journal of Scientific and Industrial Research 12A: 212-216.

Ghosh, M.K. 1956. Utilisation of solar energy. Science and Culture 22: 304-312.

Grupp, M., Montagne, P. and Wackernagel, M. 1991. A novel advanced box-type solar cooker. Solar Energy 47: 107-113.

IMD. 1985. Solar radiation atlas of India, India Meteorological Department, New Delhi

Lawand, A. T. 1973. A description of large scale steam solar cooker in Haiti, Brace research McGill Univ. Quebec, Canada.

Leon, A.M., Kumar, S. and Bhattacharya, S. C. 2002. A comprehensive procedure for performance evaluation of solar food dryers. Renewable and Sustainable Energy Review 6(4): 367–393.

Löf, G.O.G. and Fester, D. A. 1961. Design and performance of folding umbrella type solar cooker, Proceedings of U N conference on new sources of Energy, Rome, Paper S/100, 5, 347-352.

Malhotra, K.S, Nahar, N.M. and Ramana Rao B.V. 1983. Optimization factor of solar ovens. Solar Energy 31: 235-237.

MNRE. 2016. Annual Report, Ministry of New and Renewable Energy, Government of India, New Delhi.

Mullick, S. C., Kandpal, T. C. and Kumar, S. 1996. Testing of box type solar cooker: second figure of merit F_2 and its variation with load an number of pots. Solar Energy 57: 409-413.

Mullick, S. C., Kandpal, T. C. and Saxena, A. K. 1987. Thermal test procedure for box-type solar cookers. Solar Energy 39: 353-360.

Nahar N M. 2001. Design, development and testing of a double reflector hot box solar cooker with a transparent insulation material. Renewable Energy 23: 167-179.

Nahar N M. 2009. Performance and testing of a hot box storage solar cooker. Energy Conversion and Management 44: 1323–1331.

Nahar N. M., Marshall, R.H. and Brinkworth, B.J. 1994. Studies on a hot box solar cooker with transparent insulating materials. Energy Conversion and Management 35: 784-791.

Nahar, N.M. 1986. Performance studies on different models of solar cookers in arid zone conditions of India. Proc. 7th Miami International Conference on Alternative Energy Sources Hemisphere Publishing Corporation, New York 1:431-439.

Nahar, N.M. 1990. Performance and testing of an improved hot box solar cooker. Energy Conversion and Management 30: 9-16,

Olwi. I. A. and Khalifa, A. H. 1988. Computer simulation of solar pressure cooker. Solar Energy 40: 259-268

Parikh, M. and Parikh, R. 1978. Design of flat plate solar cooker for rural applications, Proc. National Solar Energy Convention of India, Bhavnagar, Central Salts and Marine Chemical Research Institute, Bhavnagar, India pp. 257-261.

Poonia, S., Singh, A.K. and Jain, D. (2018). Design development and performance evaluation of photovoltaic/thermal (PV/T) hybrid solar dryer for drying of ber (Zizyphus mauritiana) fruit. Cogent Engineering 5(1): 1-18.

Tabor, H. 1966. A solar cooker for developing countries. Solar Energy 10: 153-157.
Telkes, M. 1959. Solar cooking ovens. Solar Energy 3: 1-11.
Thanvi, K.P. and Pande, P.C. 1989. Performance evaluation of solar dryers developed at
Volunteers in Technical Assistance (VITA) 1962. Evaluation of solar cookers, Mt. Rainier. Maryland.
Von Oppen, M. 1977. The Sun basket. Appropriate Technology 4: 8-10.

7

Mathematical Model of Solar Air Heater for Performance Evaluation

Rajendra Karwa

Director, Jodhpur Institute of Engineering & Technology, Jodhpur, India

Introduction

Flat plate collector is the heart of a solar heat collection system designed for delivery of heated fluid in the low to medium temperature range (5°-70°C above ambient temperature) for applications, such as water heating, space heating, drying and similar industrial applications. The flat plate collectors absorb both beam and diffuse radiation. The absorbed radiation is converted into heat which is transferred to water or air flowing through the collector tubes or duct, respectively. Such collectors do not require tracking of the sun and little maintenance is required.

The conventional flat plate solar air heater, shown in Fig. 7.1, consists of a flat blackened absorber plate, a transparent cover (such as a glass cover) at the top and insulation at the bottom and on the sides. The air to be heated flows through the rectangular duct below the absorber plate. The glass cover transmits a major part of solar radiation incident upon it to the absorber plate where it is converted into heat. The glass is, however, opaque to long-wavelength radiation and thus it does not allow the infrared radiation from the heated absorber plate to escape.

Karwa et al. (2002) have presented the following deductions of heat collection rate and pumping power equations for the flat plate solar air heater.

The Reynolds Number ($\dot{m} D_h/\mu$) for a solar air heater duct can be expressed in a simple form as presented below considering the fact that for a rectangular duct of high aspect ratio (typically duct width W is of the order of 1 m and height H = 5-20 mm) the hydraulic diameter $D_h \approx 2H$.

$$\mathrm{Re} = \frac{\dot{m}D_h}{\mu} = \left(\frac{1}{\mu}\right)\left(\frac{AG}{WH}\right)2H = \left(\frac{1}{\mu}\right)\left(\frac{WLG}{WH}\right)2H = \frac{2GL}{\mu} \tag{1}$$

where A (= WL) is the absorber plate area, G is the mass flow rate of air per unit area of the plate and $\dot{m}$ [= $WLG/(WH)$] is the mass velocity.

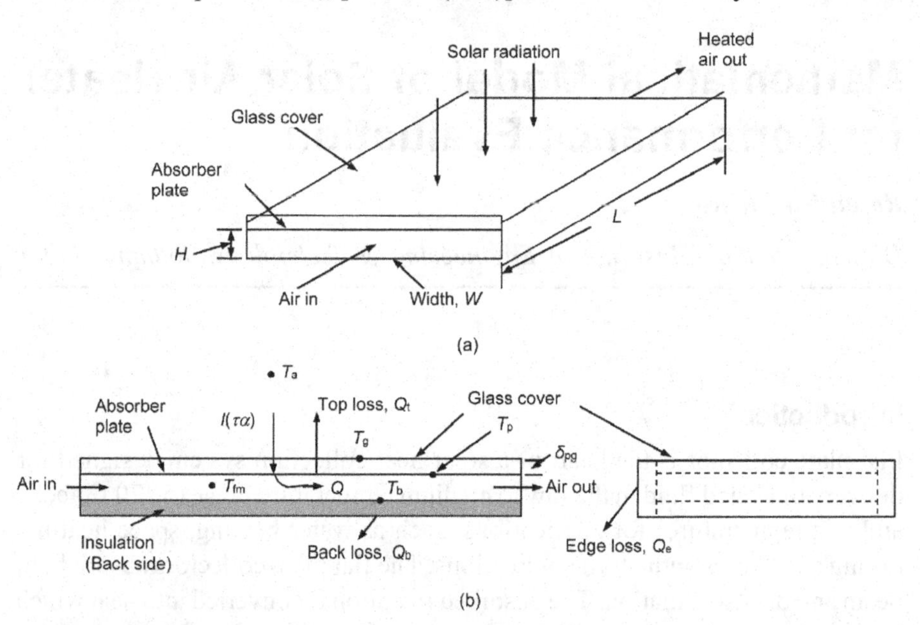

Fig. 7.1a: Schematic diagram of a solar air heater, **b** heat balance (Karwa, 2016)

Using the Dittus and Boelter correlation ($\mathrm{Nu} = 0.024\mathrm{Re}^{0.8}\mathrm{Pr}^{0.4}$) for the Nusselt number,

$$h = \frac{\mathrm{Nu}k}{D_h} = \frac{(0.024\,\mathrm{Re}^{0.8}\,\mathrm{Pr}^{0.4})k}{2H} \propto \left(\frac{GL}{\mu}\right)^{0.8}\left(\frac{1}{H}\right) \tag{2}$$

The useful heat gain Q can be expressed in the following form.

$$Q = hA\Delta T \propto (G)^{0.8}\left(\frac{L}{H}\right)\left(\frac{1}{L}\right)^{0.2} WL \tag{3}$$

Pressure loss δp and pumping power P equations for flow in the rectangular cross-section duct of a solar air heater can be written as

$$\delta p = \left(\frac{4fL}{2\rho D_h}\right)\dot{m}^2 = \left(\frac{4fL}{4\rho H}\right)\left(\frac{WLG}{WH}\right)^2 = \left(\frac{fG^2}{\rho}\right)\left(\frac{L}{H}\right)^3 \tag{4}$$

and

$$P=\left(\frac{m}{\rho}\right)\delta p=\left(\frac{WLG}{\rho}\right)\left(\frac{fG^2}{\rho}\right)\left(\frac{L}{H}\right)^3=\left(\frac{fWL}{\rho^2}\right)G^3\left(\frac{L}{H}\right)^3. \tag{5}$$

where, m (= WLG) is the mass flow rate.

Using Eq. (1) and Blasius equation for friction factor ($f = 0.0791\ Re^{-0.25}$), the pumping power equation can be transformed to give

$$P \propto G^{2.75}\left(\frac{L}{H}\right)^3 (WL)\left(\frac{1}{L}\right)^{0.25}. \tag{6}$$

From Eqs. (3) and (6), it can be seen that the heat collection rate and pumping power are strong functions of the duct geometrical parameters (L, W, H) and air flow rate per unit area of the plate G. Thus an appropriate way to evaluate the performance is to take both heat collection rate and pumping power requirement into account, i.e. to carry out a thermohydraulic performance evaluation.

Mathematical Model for Thermohydraulic Performance Prediction

Mathematical modal to predict the thermohydraulic in performance is described in Karwa et al. (2007) and Karwa and Chauhan (2010). This model calculates the useful heat gain from the iterative solution of basic heat transfer equations of top loss and equates the same with the convective heat transfer from the absorber plate to the air using proper heat transfer correlations for the smooth duct air heaters. The model estimates the collector back loss from the iterative solution of the heat balance equation for the back surface for greater accuracy. The edge loss has been calculated from the equation suggested by Klein (1975).

Fig. 7.1 shows the schematic diagram of a solar air heater and the longitudinal section of the air heater. The heat balance on a solar air heater gives the distribution of incident solar radiation, I, into useful heat gain, Q, and various heat losses, refer Fig. 7.1(b). The useful heat gain or heat collection rate can be expressed as

$$Q- = AI(\tau\alpha) - Q_L = A\,[I(\tau\alpha) - U_L(T_p - T_a)] \tag{7}$$

where,

A = Area of the absorber plate

$\tau\alpha$ = transmittance-absorptance product of the glass cover-absorber plate combination and

Q_L = Heat loss from the collector

From the known values of mean absorber plate temperature T_p and the ambient temperature T_a, overall loss coefficient U_L is calculated from

$$U_L = \frac{Q_L}{A(T_p - T_a)} \tag{8}$$

The collected heat is transferred to the air flowing through the air heater duct. Thus

$$Q = Mc_p(T_o - T_i) = GA\, c_p(T_o - T_i) \tag{9}$$

where, G (= m/A) is mass flow rate of air per unit area of the absorber plate. For open loop operation $T_i = T_a$.

Estimating the Nusselt number from appropriate correlation for the smooth duct (presented later on), heat transfer coefficient between the absorber plate and the air, h, is determined.

From heat transfer consideration, the heat gain is

$$Q = hA\,(T_p - T_{fm}) \tag{10}$$

$$\text{or } T_p = \frac{Q}{hA} + T_{fm} \tag{11}$$

where, T_{fm} is the mean temperature of air in the solar air heater duct.

The heat loss Q_L from the collector is a sum of the losses from top Q_t, back Q_b, and edge Q_e of the collector as shown in Fig. 1(b).

Top loss

The top loss, Q_t, from the collector has been calculated from iterative solution of basic heat transfer equations given below.

Heat transfer from absorber plate at mean temperature T_p to the inner surface of the glass at temperature T_{gi} takes place by radiation and convection, hence,

$$Q_{tpg} = \frac{\sigma A(T_p^4 - T_{gi}^4)}{\frac{1}{\varepsilon_p} + \frac{1}{\varepsilon_g} - 1} + h_{pg}A(T_p - T_{gi}) \tag{12a}$$

The conduction heat transfer through the glass cover of thickness $ä_g$ is given by

$$Q_{tg} = k_g A \frac{T_{gi} - T_{go}}{\delta_g} \tag{12b}$$

where, k_g is the thermal conductivity of the glass and T_{go} is temperature of the outer surface of the glass cover.

From the outer surface of the glass, the heat is rejected by radiation to the sky at tem perature T_s and by convection to the ambient, hence,

$$Q_{tgo} = \sigma A \varepsilon_g \, (T_{go}{}^4 - T_{sky}{}^4) + h_w A(T_{go} - T_\alpha) \quad (12c)$$

where, h_w is termed as wind heat transfer coefficient.

In equilibrium,

$$Q_{tpg} = Q_{tg} = Q_{tgo} = Q_t \quad (13)$$

Wind heat transfer coefficient and sky temperature

Various correlations for the estimate of the wind heat transfer coefficient h_w from the wind velocity data are available in the literature. Karwa and Chitoshiya (2013) have compiled some of these correlations and have presented a discussion on the same.

McAdams (1954) proposed the following correlation:

$$h_w = 5.6214 + 3.912 V_w \qquad \text{for } V_w \leq 4.88 \text{ m/s} \quad (14a)$$

$$= 7.172 (V_w)^{0.78} \qquad \text{for } V_w > 4.88 \text{ m/s.} \quad (14b)$$

This correlation has been widely used in modeling, simulation, and relevant calculations in spite of its shortcomings (Palyvos 2008).

The sky is considered as a blackbody at some fictitious temperature known as sky temperature T_{sky} at which it is exchanging heat by radiation. The sky temperature is a function of many parameters. Some studies assume the sky temperature to be equal to the ambient temperature because it is difficult to make a correct estimate of it, while others estimate it using different correlations. One widely used equation due to Swinbank (1963) for clear sky is

$$T_{sky} = 0.0552 (T_a)^{1.5} \quad (15)$$

where, temperatures T_{sky} and T_a are in Kelvin.

Another approximate empirical relation is (Garg and Prakash 2000)

$$T_{sky} = T_a - 6. \quad (16)$$

The above relations give significantly different values of the sky temperature.

Nowak (1989) compared and discussed the results of calculated and measured sky temperatures for a horizontal surface. They recommend that, in the case of large city areas, the sky temperature may be about 10°C higher than the one calculated from Swinbank's formula because of the atmospheric pollution.

Convective heat transfer coefficient between the absorber plate and glass cover

For the estimate of the convective heat transfer coefficient between the absorber plate and glass cover, h_{pg}, the three-region correlation of Buchberg et al. (1976) has been used, which is

$$Nu = 1 + 1.446\,(1 - 1708/Ra')^{+} \text{ for } 1708 \leq Ra' \leq 5900 \tag{17a}$$

(the + bracket goes to zero when negative)

$$Nu = 0.229\,(Ra')^{0.252} \text{ for } 5900 < Ra' \leq 9.23 \times 10^{4} \tag{17b}$$

$$Nu = 0.157\,(Ra')^{0.285} \text{ for } 9.23 \times 10^{4} < Ra' \leq 10^{6} \tag{17c}$$

where Ra' (= Ra cos β) is Rayleigh number for the inclined air layers. The Rayleigh number for the internal natural convection flow between parallel plates is given by

$$Ra = Gr\,Pr = \left[\frac{g(T_p - T_{gi})\delta_{pg}^{3}}{T_{mpg}\nu_{mpg}^{2}}\right] Pr \tag{18}$$

where δ_{pg} = gap between the absorber plate and glass cover.

The wind heat transfer coefficient, h_w, between the outer surface of the glass cover and the ambient in Eq. (12c) is a function of the wind velocity. Wind heat transfer coefficient varies from 5 $Wm^{-2}K^{-1}$ (no wind condition) to 20 $Wm^{-2}K^{-1}$ (corresponding to the usually encountered maximum wind velocity in Jodhpur).

Back and edge losses

The back loss from the collector, refer Fig. 7.1b, can be calculated from the following equation:

$$Q_b = \frac{A(T_b - T_a)}{\frac{\delta_i}{k_i} + \frac{1}{h_w}} \tag{19a}$$

where, d is the insulation thickness and k_i is the thermal conductivity of the insulating material.

Heat transfer by radiation from the heated absorber plate to the duct bottom surface Q_{pb} is given by

$$Q_{pb} = \frac{\sigma A(T_p^{4} - T_b^{4})}{\frac{1}{\varepsilon_{pi}} + \frac{1}{\varepsilon_b} - 1} \tag{19b}$$

The heat flows from the heated bottom surface at temperature T_b to the surroundings through the back insulation and to the air flowing through the duct at mean temperature T_{fm}, i.e.

$$Q_{ba} = \frac{A(T_b - T_a)}{\frac{\delta_i}{k_i} + \frac{1}{h_w}} + hA(T_b - T_{fm}) \tag{19c}$$

The heat balance for the surface gives $Q_{pb} = Q_{ba}$. The temperature of the duct bottom surface T_b can be estimated from the iterative solution of this heat balance equation.

For the edge loss estimate, the empirical equation suggested by Klein (1975) is

$$Qe = 0.5\, A_e\, (T_p - T_a) \tag{20}$$

where, A_e is the area of the edge of the air heater rejecting heat to the surroundings.

The outlet air temperature is estimated from

$$T_o = T_i + \frac{Q}{mc_p} \tag{21}$$

The thermal efficiency h of the solar air heater is the ratio of the useful heat gain Q and the incident solar radiation I on the solar air heater plane, i.e.

$$\eta = \frac{Q}{IA} \tag{22}$$

Niles et al. (1978) have used the following equations to calculate the outlet air and mean plate temperatures when the solar air heater operates in open loop mode (i.e., $T_i = T_a$):

$$T_o = T_a + \frac{I(\tau\alpha)\xi}{U_L} \tag{23}$$

$$T_p = T_i + \left[\frac{I(\tau\alpha)}{U_L}\right]\left(1 - \frac{G\xi c_p}{U_L}\right) \tag{24}$$

where, $\xi = 1 - \exp[-U_L/(Gc_p)(1 + U_L/h)^{-1}] = (F_R U_L/Gc_p)$. Parameter F_R is termed as heat removal factor and is given by (Duffie and Beckman 1980)

$$F_R = \left(\frac{Gc_p}{U_L}\right)\left[1 - \exp\left(\frac{-F'U_L}{Gc_p}\right)\right] \tag{25}$$

where, F' is termed as efficiency factor. It is given by

$$F' = \left(1 + \frac{U_L}{h}\right)^{-1} \tag{26}$$

Equations (23) and (24) may be used for the cross-check of the values of T_o and T_p calculated from Eqs. (21) and (11), respectively.

The mean air temperature equation in terms of F_R and F' (Duffie and Beckman 1980) is

$$T_{fm} = T_i + \frac{(Q/A)}{U_L F_R}\left(1 - \frac{F_R}{F'}\right) \tag{27}$$

Heat transfer and friction factor correlations

The accuracy of the results of the performance analysis strongly depends on the use of appropriate heat transfer and friction factor correlations for the solar air heater ducts. These correlations must take into account the effects of the asymmetric heating encountered in the solar air heaters, duct aspect ratio and developing length, and must be applicable to the laminar to early turbulent flow regimes. The geometry of interest is the parallel plate duct (a rectangular duct of high aspect ratio) since the width of the collector duct is of the order of 1 m and height of the order of 5-10 mm (Karwa et al. 2002, Holland and Shewen 1981), with one wall at constant heat rate and the other insulated. An intensive survey of the literature has been carried out by Karwa et al. (2007) for the correlations to fulfill these requirements.

For the asymmetrically heated high aspect ratio rectangular ducts of smooth duct solar air heater, Karwa et al. (2007) used the following correlation of Chen (in Ebadian and Dong 1998) for the apparent friction factor in the laminar regime:

$$f_{app} = \frac{24}{\mathrm{Re}} + \left(0.64 + \frac{38}{\mathrm{Re}}\right)\left(\frac{D_h}{4L}\right) \tag{28}$$

The equation takes account of the increased friction in the entrance region of the duct.

The following heat transfer correlation from Hollands and Shewen (1981) for the thermally developing laminar flow for the smooth duct can be used:

$$\mathrm{Nu} = 5.385 + 0.148\,\mathrm{Re}\left(\frac{H}{L}\right) \qquad \text{for Re} < 2550 \tag{29}$$

The friction factor correlation of Bhatti and Shah (1987) for the transition to turbulent flow regime in rectangular cross-section smooth duct ($0 \leq H/W \leq 1$) is

$$f = 1.0875 - 0.1125\left(\frac{H}{W}\right)f_o \tag{30}$$

where, $f_o = 0.0054 + 2.3 \times 10^{-8}\,\text{Re}^{1.5}$ for $2100 \le \text{Re} \le 3550$

and $f_o = 1.28 \times 10^{-3} + 0.1143\,\text{Re}^{-0.311}$ for $3550 < \text{Re} \le 10^7$

They report an uncertainty of ±5% in the predicted friction factors from the above correlation.

Considering the entrance region effect, the apparent friction factor has been determined from the following relation for flat parallel plate duct in the turbulent flow regime (Bhatti and Shah 1987):

$$f_{app} = f + 0.0175\left(\frac{D_h}{L}\right) \tag{31}$$

The Nusselt number correlations used for the transition and turbulent flow regimes from Hollands and Shewen (1981) are:

$$\text{Nu} = 4.4\times10^{-4}\,\text{Re}^{1.2} + 9.37\,\text{Re}^{0.471}\frac{H}{L} \tag{32a}$$

for $2550 \le \text{Re} \le 10^4$ (transition flow)

and

$$\text{Nu} = 0.03\,\text{Re}^{0.74} + 0.788\,\text{Re}^{0.74}\frac{H}{L} \tag{32b}$$

for $10^4 < \text{Re} \le 10^5$ (early turbulent flow)

The uncertainty of an order of 5-6% in the predicted Nusselt number can be expected (Karwa et al. 2007). As suggested by Karwa et al. (2007), the laminar regime has been assumed up to Re = 2800. The inconsistency of the predicted Nusselt number and friction factor values from correlations presented above is about 5% at the laminar-transition interface (Karwa et al. 2007).

The pressure loss, from known value of friction factor f (= f_{app} for smooth duct), and pumping power are calculated from:

$$\delta p = \left(\frac{4fL}{2\rho D_h}\right)\left(\frac{m}{WH}\right)^2 \tag{33}$$

$$P = \left(\frac{m}{\rho}\right)\delta p \tag{34}$$

Cortes and Piacentini (1990) used effective thermal efficiency η_e for the collector thermohydraulic performance evaluation, which is based on the net thermal energy collection rate of a collector considering the pumping power required to overcome the friction of the solar air heater duct. Since the power lost in overcoming frictional resistance is converted into heat, the effective efficiency equation may be defined as (Karwa and Chauhan 2010)

$$\eta_e = \frac{(Q+P) - \frac{P}{C}}{IA} \tag{35}$$

where, C is a conversion factor used for calculating equivalent thermal energy for obtaining the pumping power. It is a product of the efficiencies of the fan, electric motor, transmission, and thermo-electric conversion. For example, based on the assumption of 60% efficiency of the blower-motor combination and 33% efficiency of thermo-electric conversion process referred to the consumer point, factor C will be 0.2. Since the operating cost of a collector depends on the pumping power spent, the effective efficiency based on the net energy gain is a logical criterion for the performance evaluation of the solar air heaters.

The thermophysical properties of the air are taken at the corresponding mean temperature $T_m = T_{fm}$ or T_{mpg}. The following relations of thermophysical properties, obtained by correlating data from NBS (U.S.) (Holman 1990), can be used:

$$c_p = 1006\left(\frac{T_m}{293}\right)^{0.0155} \tag{36a}$$

$$k = 0.0257\left(\frac{T_m}{293}\right)^{0.86} \tag{36b}$$

$$\mu = 1.81\times10^{-5}\left(\frac{T_m}{293}\right)^{0.735} \tag{36c}$$

$$\rho = 1.204\left(\frac{293}{T_m}\right) \tag{36d}$$

$$\Pr = \frac{\mu c_p}{k} \tag{36e}$$

Equations (1) to (36) constitute a non-linear model for the solar air heater that can be used for the computation of the useful heat gain, Q, thermal efficiency, η, pressure loss, δp, pumping power, P, and effective efficiency, η_e.

The model has been solved by Karwa and Baghel (2014) following an iterative process depicted in Fig. 7.2. For the estimate of heat collection rate, Karwa and Chauhan (2010) and Karwa and Baghel (2014) terminated the iteration when the successive values of the plate and mean air temperatures differed by less than 0.05 K. The iteration for the estimate of top loss by them has been continued till the heat loss estimates from the absorber plate to the glass cover and glass cover to the ambient, i.e., Q_{tpg} and Q_{tgo} from Eqs. (12a) and (12c), respectively, differed by less than 0.2%.

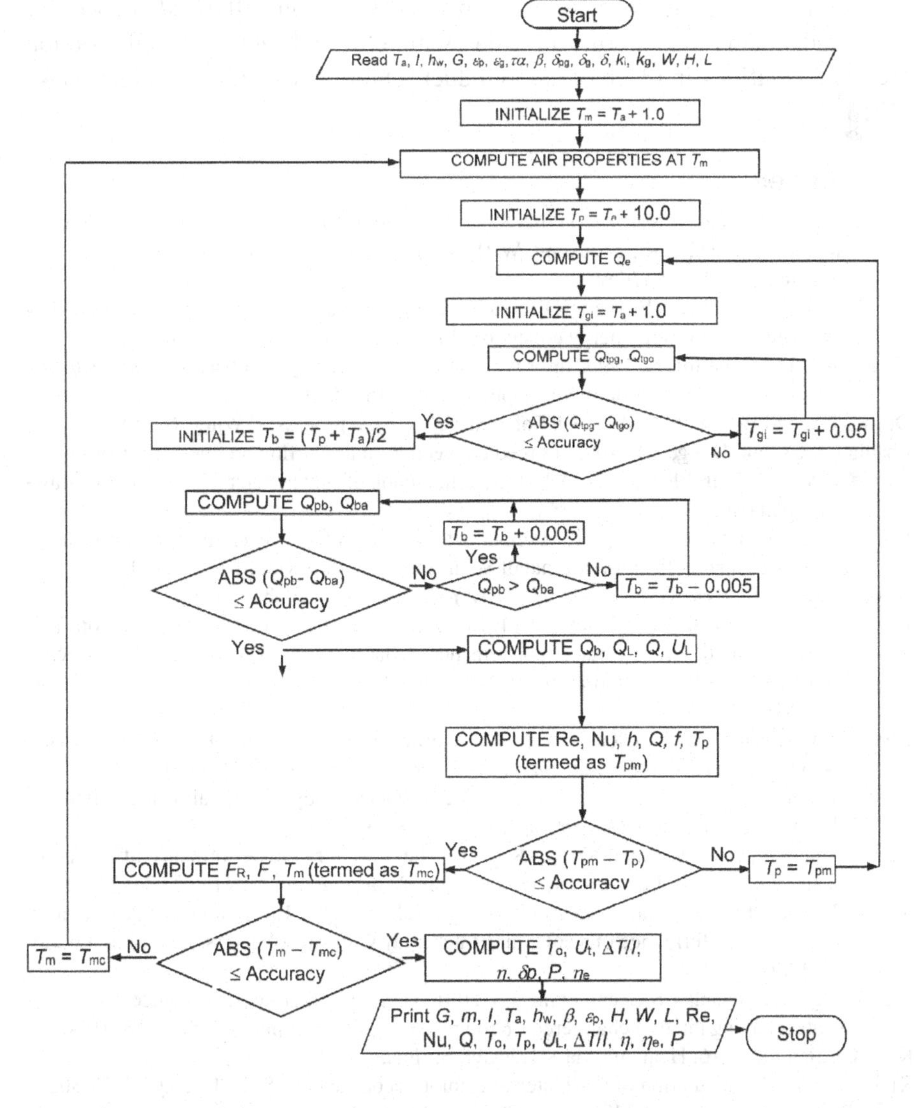

Fig. 7.2: Flow chart for iterative solution of mathematical model (Karwa and Baghel)

The mathematical model presented here has been validated by Karwa et al. (2007) against the data from the experimental study of a smooth duct solar air heater of Karwa et al. (2001) with reported uncertainties of ±4.65% in Nusselt number and ±4.56% in friction factor. The standard deviations of the predicted values of thermal efficiency and pumping power from the experimental values of these parameters from (Karwa et al. 2001) have been reported by Karwa et al. (2007) to be ±4.9% and ±6.2%, respectively.

Enhanced performance solar air heaters have been developed using artificial roughness on the absorber plate. Karwa and Chauhan (2010) also used the presented model for such solar air heater with relevant heat transfer and friction factor correlations in place of smooth duct relations. For details, refer Karwa (2016).

References

Bhatti, M.S. and Shah, R.K.1987. Turbulent and transition flow convective heat transfer. In: Kakac, S., Shah, R.K., and Aung, W. (Eds.), Handbook of single-phase convective heat transfer. New York: Wiley.

Buchberg, H., Catton, I. and Edwards, D.K. 1976. Natural convection in enclosed spaces-a review of application to solar energy collection. Journal of Heat Transfer 182-8.

Cortes, A. and Piacentini, R. 1990. Improvement of the efficiency of a bare solar collector by means of turbulence promoters. Applied Energy 36: 253-61.

Duffie, J.A. and Beckman, W.A. 1980. Solar energy thermal processes. Wiley, New York.

Ebadian, M.A. and Dong, Z.F. 1998. Forced convection, internal flow in ducts. In: Rohsenow, W.M., Hartnett, J.P. and Cho, Y.I. (Eds.), Handbook of heat transfer. New York: McGraw-Hill; [Chapter 5].

Hollands, K.G.T. and Shewen, E.C. 1981. Optimization of flow passage geometry for air-heating, plate-type solar collectors. Journal of Solar Energy Engineering. 103: 323-30.

Holman, J.P. 1990. Heat Transfer. 7th ed., New York: McGraw-Hill book Co.

Karwa, Rajendra and Baghel, Shweta. 2014. Effect of parametric uncertainties, variations and tolerances on thermo-hydraulic performance of flat plate solar air heater. J Solar Energy Hindawi Publishing Corporation 2014, Article ID 194764, 18 pages. doi: 10.1155/2014/194764

Karwa, Rajendra and Chauhan, Kalpana. 2010. Performance evaluation of solar air heaters having v-down discrete rib roughness on the absorber plate. Energy 35: 398-409.

Karwa, Rajendra and Chitoshiya, Girish. 2013. Performance study of solar air heater having v-down discrete ribs on absorber plate. Energy 55: 939-955.

Karwa, Rajendra, Garg, S.N. and Arya, A.K. 2002. Thermo-hydraulic performance of a solar air heater with n-subcollectors in series and parallel configuration. Energy 27: 807-12.

Karwa, Rajendra, Karwa, Nitin, Misra, Rohit and Agarwal, P.C. 2007. Effect of flow maldistribution on thermal performance of a solar air heater array with subcollectors in parallel. Energy 32: 1260-70.

Karwa, Rajendra, Solanki, S.C. and Saini, J.S. 2001. Thermo-hydraulic performance of solar air heaters having integral chamfered rib roughness on absorber plates. Energy 26: 161-176.

Karwa, Rajendra. 2016. Heat and Mass Transfer, Springer.

Klein, S.A. 1975. Calculation of flat-plate collector loss coefficients. Solar Energy 17: 79-80.

Niles, P.W., Carnegie, E.J., Pohl, J.G. and Cherne, J.M. 1978. Design and performance of an air collector for industrial crop dehydration. Solar Energy 20: 19-23.

Nowak, H .1989. The sky temperature in net radiant heat loss calculations from low-sloped roofs. Infrared Physics 29 (2-4)): 231-32.
Palyvos, J.A. 2008. A survey of wind convection coefficient correlations for building envelope energy systems' modeling. Applied Thermal Engineering 28(8-9): 801-808.
Swinbank, W.C.1963. Long-wave radiation from clear skies. Q J Roy Meteor. Soc 89: 339.

Nomenclature

A	absorber plate area = WL, m^2
D_h	hydraulic diameter of duct = $4WH/[2(W+H)]$, m
f	fanning friction factor
G	mass flow rate per unit area of plate = m/A, kg s^{-1} m^{-2}
H	air flow duct height (depth), m
h_w	wind heat transfer coefficient, Wm^{-2} K^{-1}
I	solar radiation on the collector plane, Wm^{-2}
k	thermal conductivity of air, Wm^{-1}K^{-1}
L	length of collector, m
m	mass flow rate, kgs^{-1}
Nu	nusselt number
P	pumping power, W
Pr	prandtl number = $ìc_p/k$
Q	useful heat gain, W
Re	reynolds number = $[m/(WH)]D_h/ì$
T_a	ambient temperature, K
T_{fm}, T_m	mean air temperature = $(T_o + T_i)/2$, K
T_i	inlet air temperature, K
T_{mpg-}	mean of the plate and glass temperatures = $(T_p + T_{gi})/2$, K
T_o	outlet air temperature, K
T_p	mean plate temperature, K
T_{sky}	sky temperature, K
U_L	overall loss coefficient, Wm^{-2} K^{-1}
W	width of the duct, m

Greek symbols

β	collector slope, deg
δ_{pg}	gap between the absorber plate and glass cover, m
ΔT	air temperature rise = $T_o - T_i$, K

ε emissivity

η thermal efficiency

η_e effective efficiency

μ dynamic viscosity, Pa s

ν_{mpg} *kinematic viscosity of air at temperature* T_{mpg}, *m^2/s*

$\tau\alpha$ transmittance-absorptance product

Subscripts

b duct bottom surface

g glass

m mean

p plate

8

Basic Principles of Flat Plate and Evacuated Tube Collectors in Solar Water Heater

Surendra Poonia, A.K. Singh, P. Santra and R.K. Singh

ICAR-Central Arid Zone Research Institute, Jodhpur, Rajasthan, India

Introduction

We are blessed with solar energy in abundance at no cost. The solar radiation incident on the surface of the earth can be conveniently utilized for the benefit of human society. Solar energy can be used as thermal energy for water heating, cooking, drying, distillation, space heating, cooling and power generation or it can be converted to electricity through photovoltaic cells, commonly known as solar cells. One of the popular devices that harness the solar energy is solar hot water system (SHWS). Flat plate collectors are used for low temperature applications (below 100°C) while concentrators are preferred for higher temperatures. The most efficient application of flat plate collector is for getting hot water. Hot water in winter is an essential requirement for domestic uses such as bathing, cleaning of utensils and washing of clothes. In rural areas hot water is also required for softening of animal feed. Generally it is obtained by using firewood and cow dung cake in rural areas or using kerosene, liquid petroleum gas, coal or electricity in urban areas. Therefore, the use of solar water heaters will conserve lot of commercial and non-commercial fuels which are being wasted in merely getting hot water. A solar water heating system (SWHS) is made of several important elements: one or more solar collectors, a pump, a heat exchanger, a storage tank (or multiple tanks) and a back-up storage tank. The Solar heating system can be classified as passive or active. For water heating purposes, the general practice is to use flat plate solar energy collectors (FPC). The evacuated tube collectors (ETC) and evacuated tube heat pipe collectors (ETHP) are more efficient, though, the initial cost is comparatively higher. There are three type of water heaters extensively studied by various workers *viz*. natural circulation, collector-cum-storage and forced circulation in all over the world.

Solar Collectors

At the heart of a solar thermal system is the solar collector. It absorbs solar radiation, converts it into heat, and transfers useful heat to the solar system. There are a number of different design concepts for collectors: besides simple absorbers used for swimming pool heating, more sophisticated systems have also been developed for higher temperatures, such as integral storage collector systems, flat plate collectors, evacuated flat plate collectors and evacuated tube collectors.

A solar collector is very efficient at turning sun light into heat.There is a special coating called an "Absorber Surface Coating" that is spluttered or fused at a very high temperature, with the metal sheet inside the solar collector. This surface makes the collector efficient and effective. It is all about the "aperture area or absorber surface" which is part of the collector that collects light and transforms into heat. All efficient panels are made of a strong aluminum body that is non-corrosive. The heat absorbed by the solar collector is protected by a toughened solar glass that covers the entire solar panel. This transparent glass cover prevents wind and breezes from carrying the collected heat away (convection). Together with the frame, the glass protects the absorber from adverse weather conditions. Typical frame materials include aluminum and galvanized steel; sometimes fiberglass-reinforced plastic is used. As far as the water heating system is concerned, the panels may be either flat plate type or vacuum tube type. Both flat plate and vacuum tube collectors can work on cloudy days. If we compare flat plates and evacuated tubes by aperture area, vacuum tubes are more efficient than the flat plates.

Concept of solar water heating system

A solar water heater consists of a collector to collect solar energy and an insulated storage tank to store hot water. The solar energy incident on the absorber panel coated with selected coating transfers the heat to the riser pipes underneath the absorber panel. The water passing through the risers gets heated up and is delivered into the storage tank. There circulation of the same water through absorber panel in the collector raises the temperature to 80°C (maximum) in a good sunny day. The total system with solar collector, storage tank and pipelines is called solar hot water system.

Broadly, the solar water heating systems are of two categories. They are: closed loop system and open loop system. In the first one, heat exchangers are installed to protect the system from hard water obtained from bore wells or from freezing temperatures in the cold regions. In the other type, either thermo-syphon or forced circulation system, the water in the system is open to the atmosphere at one point or other. The thermo-syphon systems are simple and relatively inexpensive. They are suitable for domestic and small institutional systems, provided the water is treated and potable in quality. The forced circulation systems employ electrical pumps to circulate the water through collectors and storage tanks.

The choice of system depends on heat requirement, weather conditions, heat transfer fluid quality, space availability, annual solar radiation, etc. The SHW systems are economical, pollution free and easy for operation in warm countries like ours.

Based on the collector system, solar water heaters can be of two types:

1) Flat plate collector based SWHS (FPC-SWHS)
2) Evacuated tube collector based SWHS (ETC-SWHS).

For water heating purposes, the general practice is to use Flat plate solar collectors (FPC). But the evacuated tube heat pipe collectors or evacuated tube collectors (ETHP or ETC) are more efficient, even the initial cost is comparatively high.

Flat-plate collectors

The majority of solar collectors that are sold in many countries are of the flat plate variety. The FPC-SWHS, can generate hot water between 60°C and 70°C, depending on the size and quality of the collectors. A flat plate collector consists of an absorber, a transparent cover, a frame, and insulation. Usually, a solar safety glass is used as a transparent cover, as it transmits a great amount of short-wave light spectrum. The absorber, inside the flat plate collector converts sunlight to heat and transfers it to water in the absorber tubes. The absorber is usually made of metal materials such as copper, steel or aluminum. The collector can be made of plastic, metal or wood and the glass front cover must be sealed so that heat does not escape, and dirt, insects or humidity do not get into the collector itself. Absorbers are usually black, as dark surfaces will have high degree of light absorption. As the absorber warms up to a temperature higher than the ambient temperature, it gives off a great part of the accumulated solar energy in the form of long-wave heat rays. The ratio of absorbed energy to emitted heat is indicated by the degree of emission. The most efficient absorbers have a selective surface coating, which enables the conversion of a high proportion of solar radiation into heat, simultaneously reducing the emission of heat. The usual coatings provide a degree of absorption of over 90%. Solar paints, which are painted or sprayed manually on the absorbers, are not very selective, as they have a high level of emission. Some selective coatings that include black chrome, black nickel and aluminum oxide with nickel will absorb the heat more efficiently. Relatively new is a titanium-nitride-oxide layer, which is applied via steam in a vacuum process. This type of coating stands out because of its low emission rates and efficiency. The efficiency of the flat plate collectors varies from 40 to 70% depending upon operation temperature. Fig.8.1 & 8.2 shows the processes occurring at a flat plate collector. There are 60 BIS approved manufacturers of solar flat plate collectors in our country.

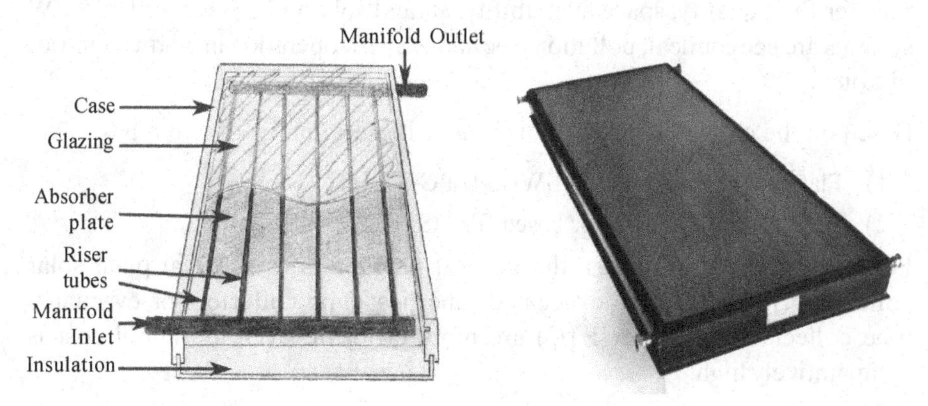

Fig. 8.1: Flat plate collector

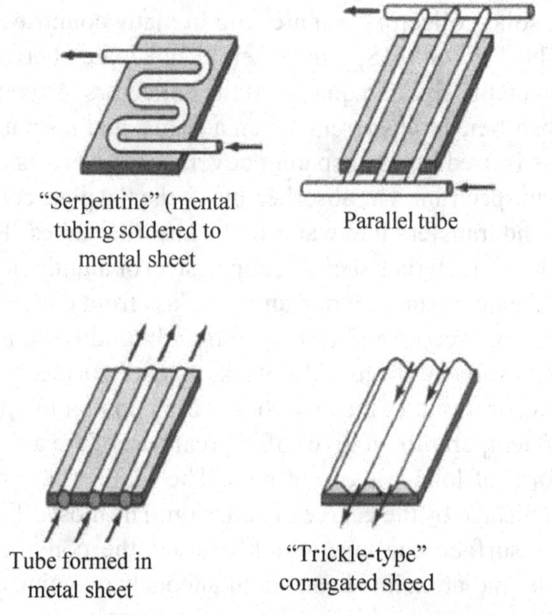

Fig. 8.2: Types of solar flat plate collectors

Evacuated Tube Collector

In order to reduce heat loss within the frame by convection, the air can be pumped out of the collector tubes. Such collectors then can be called as evacuated-tube collectors (Fig. 8.3). They must be re-evacuated once every one to three years. Evacuated tube solar collectors are very efficient and can achieve very high temperatures. They are well suited to commercial and industrial heating applications and can be an effective alternative to flat plate collectors for industrial purposes; especially in areas where it is often cloudy. An evacuated-tube collector contains several rows of glass tubes connected to a header pipe. Each tube has the air removed from it to eliminate heat loss through convection and radiation. Evacuated tube collector is made of double layer borosilicate glass tubes evacuated for providing insulation. The outer wall of the inner tube is coated with selective absorbing material. This helps absorption of solar radiation and transfers the heat to the water which flows through the inner tube. There are 44 MNRE approved ETC based solar water heating suppliers in India. Evacuated collectors are good for applications requiring energy delivery at moderate to high temperatures (domestic hot water, space heating and process heating applications typically at 60°C to 80°C depending on outside temperature), particularly in cold climates. A direct flow evacuated tube collector consists of a glass tube, with a flat or curved aluminum fin attached to a metal or glass pipe. The fin is covered with a selective coating that transfers heat to the fluid that is circulating through the pipes, one for inlet fluid and the other for outlet fluid. In this type of vacuum collector, the absorber strip is located in an evacuated and pressure proof glass tube. The heat transfer fluid flows through the absorber directly in a U-tube orin countercurrent in a tube-in-tube system.

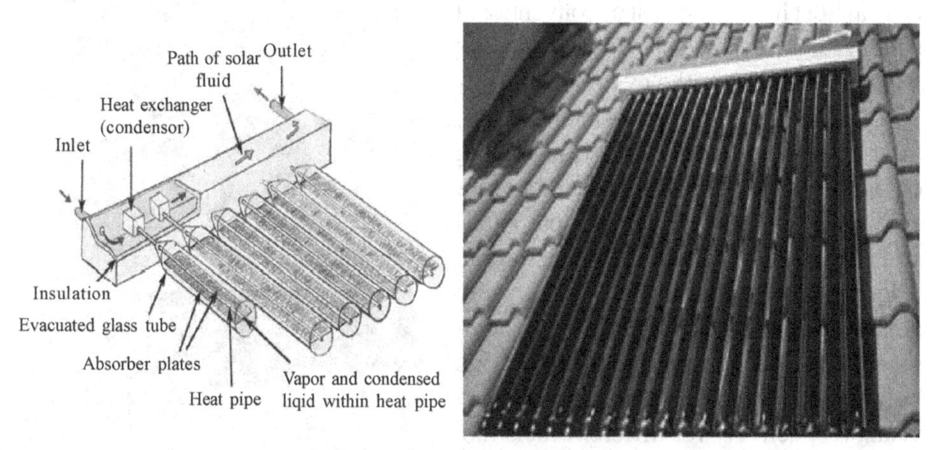

Fig. 8.3: Evacuated tube collector

Natural circulation and collector-cum-storage solar water heater

There are generally three types of solar water heaters extensively studied by various workers *viz.* natural circulation, collector-cum-storage and forced circulation. The natural circulation type solar water heaters were studied extensively all over the world. Most suitable solar water heaters for domestic purpose are the natural circulation and collector-cum-storage type which have been designed and developed at the ICAR-Central Arid Zone Research Institute, Jodhpur. In commercially available natural circulation type solar water heater mainly copper pipes and copper sheet are used for flat plate collectors. Nahar (1984, 1988 and 1992) found that flat plate collector using GI pipes header and riser and aluminum sheet as absorber saves 30% cost while giving better performance comparable to commercially available copper pipes and copper sheet flat plate collectors (Fig. 8.4). The heater can provide 100 litre hot water at 60-70°C in the evening, which can be retained to 50-60°C till next day. Such flat plate collectors have been installed in hotels, hostels, guest-houses etc. These water heaters are also suitable for agro-based industries.

Fig. 8.4: Natural circulation type solar water heater

On the other hand, collector-cum-storage solar water heater (Fig. 8.5) reduces the cost, almost half of the cost of conventional solar water heater, and provide 100 litre hot water at 50-60°C in the evenings and 40-45°C next day mornings (winter season) after covering the device with insulating cover. Such type of water heaters were studied in detail (Nahar and Gupta 1988) and installed at villages for demonstration. The payback period of 100 litre capacity natural circulation type and collector-cum-storage type solar water heater varies between 1.6 to 10.8 years and 1.1 to 6.5 years, respectively for different fuels like firewood, coal, electricity, LPG and kerosene.

Fig. 8.5: Collector cum storage type solar water heater

BIS standards of flat plate collectors are available and standards for storage tanks are being developed. Tubular collectors are getting more popular due to low cost. Use of large size solar water heaters have been demonstrated in dairy and textile industries in Maharashtra, Karnataka and Gujarat states.

Salient features of solar water heating system

i) Solar hot water system turns cold water into hot water with the help of sun's rays.

ii) Around 60–80°C temperature can be attained depending on solar radiation, weather conditions and solar collector system efficiency.

iii) Hot water for homes, hostels, hotels, hospitals, restaurants, dairies, industries etc.

iv) Can be installed on roof-tops, building terrace and open ground where there is no shading, south orientation of collectors and over-head tank above SWH system.

v) SWH system generates hot water on clear sunny days (maximum), partially clouded (moderate) but not in rainy or heavy overcast day.

vi) Only soft and potable water can be used.

vii) Stainless Steel is used for small tanks whereas Mild Steel tanks with anticorrosion coating inside are used for large tanks.

viii) Solar water heaters (SWHs) of 100-300 litres capacity are suited for domestic application, while, larger systems can be used in restaurants, guest houses, hotels, hospitals, industries etc.

Conclusion

The Flat plate collector can generate hot water between 60-70°C, depending on the size and quality of the collectors, evacuated tube solar collectors are very efficient and can achieve very high temperatures. A 100 liters capacity SWH can replace an electric geyser for residential use and saves 1500 units of electricity annually and prevent emission of 1.5 tonnes of carbon dioxide per year. Though the initial cost for procuring the SWHS is high, the returns on investment can be achieved within a short period by means of saving the firewood. The payback period depends on the site of installation, utilization pattern and fuel replaced. Typically, the solar thermal systems will reduce the firewood consumption, deforestation and pollution.

References

Nahar, N.M. 1984. Energy conservation and field performance of a natural circulation type solar water heater. Energy 5: 461-464

Nahar, N.M. 1988. Performance and testing of a low cost solar water heater-cum-solar cooker. Solar and Wind Technology 5: 611-615.

Nahar, N.M. 1992. Energy conservation and payback periods of natural circulation type solar water heaters. International Journal of Energy Research 16: 445-452.
Nahar, N.M. and Gupta, J.P. 1988. Studies on collector-cum-storage type solar water heaters under arid zone conditions of India. International Journal of Energy Research 12: 147-153.

9

Basic Principles of Solar Refrigeration and its Applications

Priyabrata Santra, S. Poonia and R.K. Singh

ICAR-Central Arid Zone Research Institute, Jodhpur, Rajasthan-342003

Introduction

Solar refrigeration technology engages a system where solar power is used for cooling purposes. It offers a wide variety of cooling techniques powered by solar collector-based thermally driven cycles and photovoltaic (PV)-based electrical cooling systems. Cooling can be achieved through four basic methods: solar PV cooling, solar thermal cooling, solar thermo-electrical cooling and solar thermo-mechanical cooling. The first is a PV-based solar energy system, where solar energy is converted into electrical energy and used for refrigeration much like conventional methods. The second method utilizes a solar thermal refrigeration system, where a solar collector directly heats the refrigerant through collector tubes instead of using solar electric power. The third one produce cool by thermoelectric processes. The fourth method converts the thermal energy to mechanical energy, which is utilized to produce the refrigeration effect.

The performance of solar refrigeration systems is determined based on energy indicators of these systems. The COP (coefficient of performance) can be calculated as follows:

$$\text{COP} = \frac{Q_e}{Q_s} \quad (1)$$

where Q_e is the cooling power; Q_s is the consumed solar energy by the system. In a solar refrigeration system, COP is determined by individual efficiency of solar system e.g. PV, thermal, thermo-electrical, thermomechanical etc and the efficiency of refrigeration process. For example, if the efficiency of PV module is 10% and COP for refrigeration process is 3.0, the COP of the solar PV refrigeration system will be 30%. Apart from COP, the performance of solar refrigeration system is also defined by energy efficiency ratio (EER), in British thermal unit per Watt-hours (Btu/(wh)), which is calculated as follows:

$$EER = 3.413 \times COP \quad (2)$$

In the following sections, basic four methods of solar refrigeration are discussed.

Solar photovoltaic cooling systems

A PV cell is basically a solid-state semiconductor device that converts light energy into electrical energy. To accommodate the huge demand for electricity, PV-based electricity generation has been rapidly increasing around the world alongside conventional power plants over the past two decades. Fig. 9.1 shows a representation of the development of solar PV cooling systems. While the output of a PV cell is typically direct current (DC) electricity, most domestic and industrial electrical appliances use alternating current (AC). Therefore, a complete PV cooling system typically consists of four basic components: photovoltaic modules, a battery, an inverter circuit and a vapour compression AC unit.

- The PV cells produce electricity by converting light energy (from the sun) into DC electrical energy.
- The battery is used for storing DC voltages at a charging mode when sunlight is available and supplying DC electrical energy in a discharging mode in the absence of daylight. A battery charge regulator can be used to protect the battery form overcharging.
- The inverter is an electrical circuit that converts the DC electrical power into AC and then delivers the electrical energy to the AC loads.
- The vapour compression AC unit is actually a conventional cooling or refrigeration system that is run by the power received from the inverter.

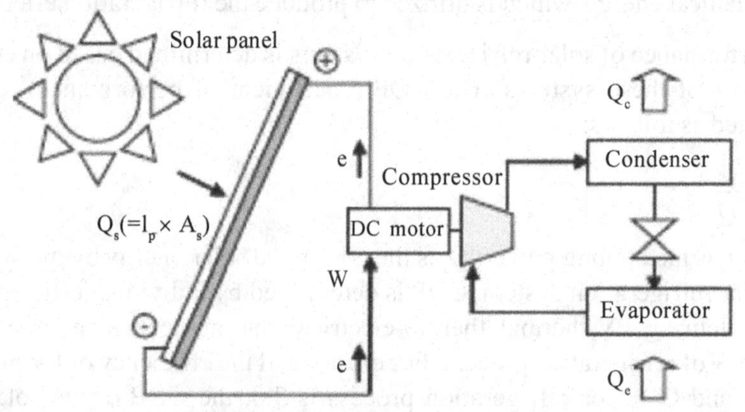

Fig. 9.1: Schematic of a solar PV cooling system

Solar thermo-electrical cooling

In solar electric cooling, power produced by the solar PV devices is supplied to the Peltier cooling systems. In the thermoelectric processes the cool is produced using the principle of producing electricity from solar energy through thermoelectric effect and the principle of producing cool by Peltier effect. The principle diagram of thermoelectric refrigeration is given in Fig. 9.2. Thermoelectric generator consists of a small number of thermocouples that produce a low thermoelectric power but which can easily produce a high electric current. It has the advantage that it can be operated with a low level heat source and is therefore useful to convert solar energy into electricity. The thermoelectric refrigerator is also composed of a small number of thermocouples through which the current is produced by the generator. The combination of the two parts is compatible with use as thermoelectric materials of the semiconductors based on Bi_2Te_3. A thermoelectric generator, which draws its heat from solar energy, is a particularly suitable source of electrical power for the operation of a thermoelectric refrigerator.

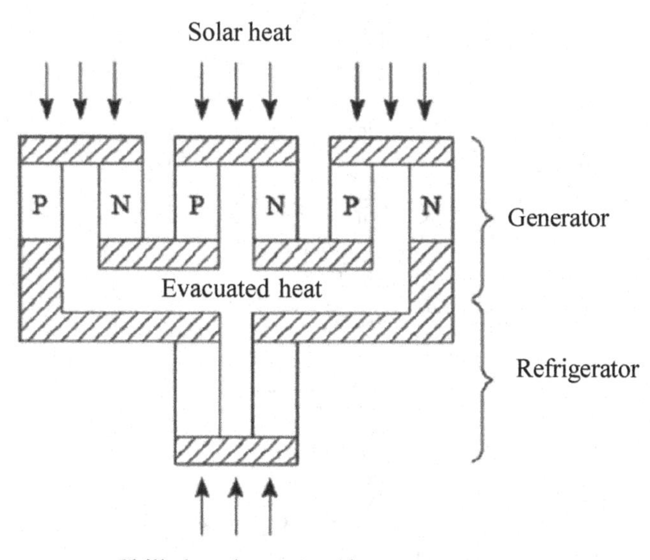

Fig. 9.2: Schematic of thermo-electric cooling system

The thermoelectric refrigerator is a unique cooling system, in which the electron gas serves as the working fluid. In recent years, concerns of environmental pollution due to the use of CFCs in conventional domestic refrigeration systems have encouraged increasing activities in research and development of domestic refrigerators using Peltier modules. Moreover, recent progress in thermoelectric and related fields have led to significant reductions in fabrication costs of Peltier modules and heat exchangers together with moderate improvements in the module performance. Although the COP of a Peltier module is lower than that of conventional compressor unit, efforts have been made to develop domestic thermoelectric cooling systems to exploit the advantages associated with this solid-state energy conversion technology.

Thermo-mechanical refrigeration

In a solar thermo-mechanical refrigeration system, a heat engine converts solar heat to mechanical work, which in turn drives a mechanical compressor of a vapour compression refrigeration machine. A schematic diagram of such a cooling system is shown in Fig. 9.3.

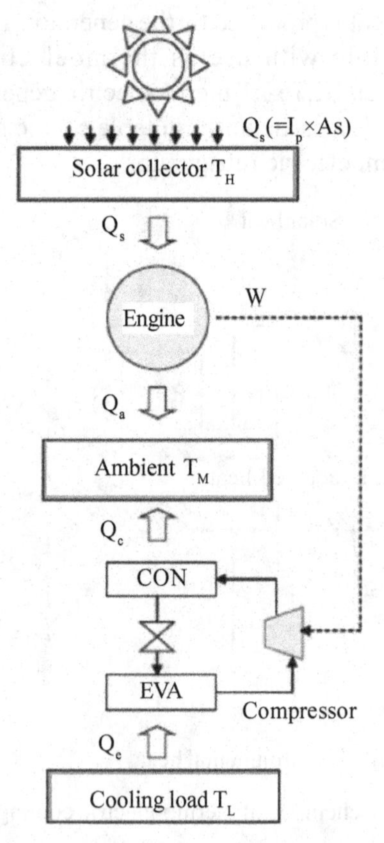

Fig. 9.3: Solar thermo-mechanical refrigeration system

In the figure, a solar collector receives solar radiation Q_s [the surface area A_s (m^2) multiplied by the solar radiation perpendicular to the surface I_p (kW m^{-2})] from the sun and supplies Q_g to a heat engine at the temperature T_H. The ratio of supply heat Q_g to the radiation Q_s is defined as the thermal efficiency of a solar thermal collector, $\eta_{sol\text{-}heat}$.

$$\eta_{\text{sol.heat}} = \frac{Q_g}{I_p \times A_s} = \frac{Q_g}{Q_s} \tag{3}$$

$\eta_{\text{sol-heat}}$ is less than 1 due to optical and thermal losses.

A heat engine produces mechanical work W and rejects heat Q_a to ambient at temperature T_M. The efficiency of engine, $\eta_{\text{heat-pow}}$ is defined as the work produced per heat input Q_g as follows.

$$\eta_{\text{heat.pow}} = \frac{W}{Q_g} \tag{4}$$

The mechanical work W in turn drives the compressor of the refrigeration machine to remove heat Q_e from the cooling load at temperature T_L. Waste heat Q_c, which is equal to the sum of Q_e and W, is rejected to ambient at the temperature T_M. Efficiency of the refrigeration machine is given in the following equation.

$$\eta_{\text{pow.cool}} = \frac{Q_e}{W} \tag{5}$$

Then the overall efficiency of a solar thermo-mechanical refrigeration system is given by the three efficiencies in Eqs. (3), (4) and (5) as follows:

$$\eta_{\text{sol.cool}} = \eta_{\text{sol.heat}} \times \eta_{\text{heat.pow}} \times \eta_{\text{pow.cool}} = \frac{Q_e}{Q_s} \tag{6}$$

Solar thermal cooling techniques

Solar thermal cooling is becoming more popular because a thermal solar collector directly converts light into heat. Fig. 9.4 shows a schematic diagram of a solar thermal cooling system. The solar collection and storage system consists of a solar collector (SC) connected through pipes to the heat storage. Solar collectors transform solar radiation into heat and transfer that heat to the heat transfer fluid in the collector. The fluid is then stored in a thermal storage tank (ST) to be subsequently utilized for various applications. The thermal AC (air-conditioning) unit is run by the hot refrigerant coming from the storage tank, and the refrigerant circulates through the entire system.

Sorption technology is utilized in thermal cooling techniques. Sorption refrigeration uses physical or chemical attraction between a pair of substances to produce refrigeration effect. A sorption system has a unique capability of transforming thermal energy directly into cooling power. Among the pair of substances, the substance with lower boiling temperature is called sorbate and the other is called sorbent. The sorbate plays the role of refrigerant. The cooling effect is obtained from the chemical or physical changes between the sorbent and the refrigerant. Sorption technology can be classified either as open sorption systems or closed sorption systems.

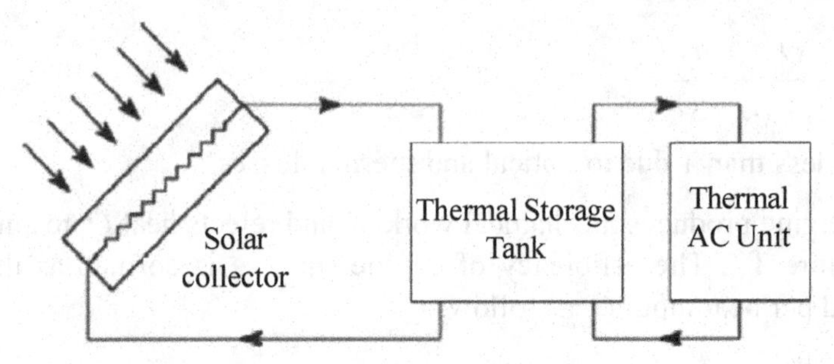

Fig. 9.4: Schematic of a solar thermal cooling system

Open sorption systems

Open sorption cooling is more commonly called desiccant cooling because sorbent is used to dehumidify air. Various desiccants are available in liquid or solid phases. Basically all water absorbing sorbents can be used as a desiccant. Examples are silica gel, activated alumina, zeolite, LiCl and LiBr.

In a liquid desiccant cooling system, the liquid desiccant circulates between an absorber and a regenerator in the same way as in an absorption system. Main difference is that the equilibrium temperature of a liquid desiccant is determined not by the total pressure but by the partial pressure of water in the humid air to which the solution is exposed to. A typical liquid desiccant system is shown in Fig. 9.5. In the dehumidifier of Fig. 9.5, a concentrated solution is sprayed at point A over the cooling coil at point B while ambient or return air at point 1 is blown across the stream. The solution absorbs moisture from the air and is simultaneously cooled down by the cooling coil. The results of this process are the cool dry air at point 2 and the diluted solution at point C. Eventually an aftercooler cools this air stream further down. In the regenerator, the diluted solution from the dehumidifier is sprayed over the heating coil at point E that is

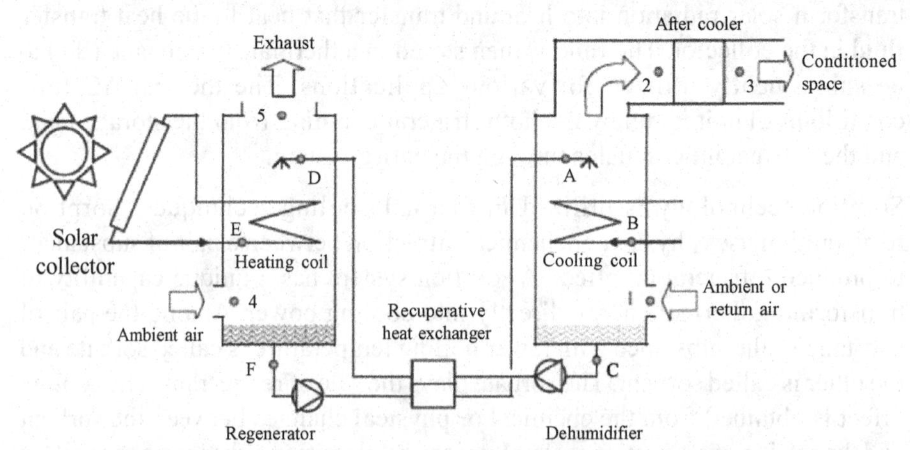

Fig. 9.5: A liquid desiccant cooling system with solar collector

connected to solar collectors and the ambient air at point 4 is blown across the solution stream. Some water is taken away from the diluted solution by the air while the solution is being heated by the heating coil. The resulting concentrated solution is collected at point F and hot humid air is rejected to the ambient at point 5. A recuperative heat exchanger preheats the cool diluted solution from the dehumidifier using the waste heat of the hot concentrated solution from the regenerator, resulting in a higher COP.

A solid desiccant cooling system is quite different in its construction mainly due to its non-fluid desiccant. Fig. 9.6 shows an example of a solar-driven solid desiccant cooling system. The system has two slowly revolving wheels and several other components between the two air streams from and to a conditioned space. The return air from the conditioned space first goes through a direct evaporative cooler and enters the heat exchange wheel with a reduced temperature (A/B). It cools down a segment of the heat exchange wheel which it passes through (B/C). This resulting warm and humid air stream is further heated to an elevated temperature by the solar heat in the heating coil (C/D). The resulting hot and humid air regenerates the desiccant wheel and is rejected to ambient (D/E). On the other side, fresh air from ambient enters the regenerated part of desiccant wheel (1/2). Dry and hot air comes out of the wheel as the result of dehumidification. This air is cooled down by the heat exchange wheel to a certain temperature (2/3). Depending on the temperature level, it is directly supplied to the conditioned space or further cooled in an aftercooler (3/4). If no aftercooler is used, cooling effect is created only by the heat exchange wheel, which was previously cooled by the humid return air at

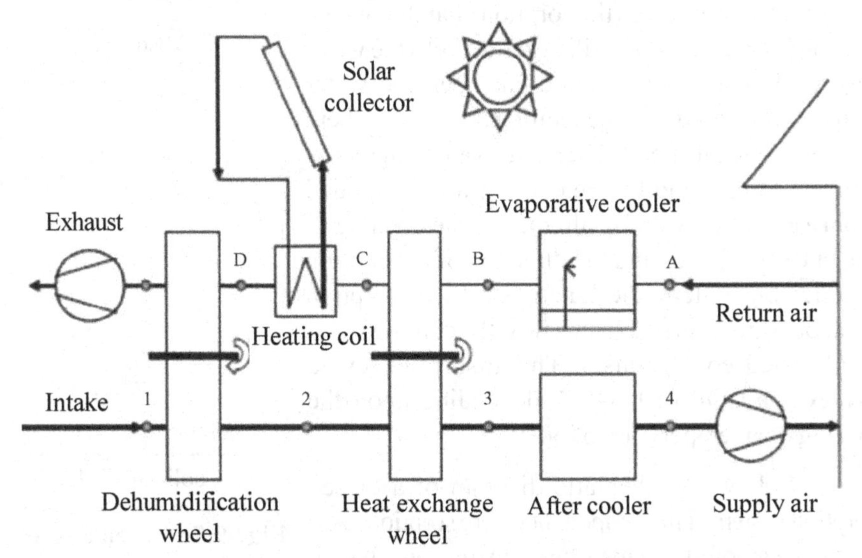

Fig. 9.6: A solid desiccant cooling system with solar collector.

point B on the other side. Temperature at point 3, T3, cannot be lower than TB, which in turn is a function of the return air condition at point A.

Closed sorption systems: In closed sorption technology, there are two basic methods: absorption refrigeration and adsorption refrigeration. Absorption is the process in which a substance assimilates from one state into a different state. These two states create a strong attraction to make a strong solution or mixture. The absorption refrigeration technology consists of a generator, a pump and an absorber that are collectively capable of compressing the refrigerant vapour. The evaporator draws the vapour refrigerant by absorption into the absorber. The extra thermal energy separates the refrigerant vapour from the rich-solution. The condenser condenses the refrigerant by rejecting the heat and then the cooled liquid refrigerant is expanded by the evaporator, and the cycle is completed. Typical refrigerant/absorbent pairs used in the absorption system are NH_3/H_2O and H_2O/LiBr. Solar absorption cooling systems utilize the thermal energy from a solar collector to separate a refrigerant from the refrigerant/absorbent mixture. The adsorption process differs from the absorption process in that absorption is a volumetric phenomenon, whereas adsorption is a surface phenomenon. The primary component of an adsorption system is a solid porous surface with a large surface area and a large adsorptive capacity. The adsorption cycle is composed of two sorption chambers, an evaporator, and a condenser. Water is vaporized under low pressure and low temperature in the evaporator. Then the water vapour enters the sorption chamber where the solid sorbent, such as silica gel, adsorbs the water vapour. In the other sorption chamber, the water vapour is released by regenerating the solid sorbent by applying the heat. Then the water vapour is condensed to liquid by the cooling water supplied from a cooling tower. By alternating the opening of the butterfly valves and the direction of the cooling and heating waters, the functions of two sorption chambers are reversed. In this way, the chilling water is obtained continuously. The adsorption cycle achieves a COP of 0.3–0.7, depending upon the driving heat temperature of 60–95°C.

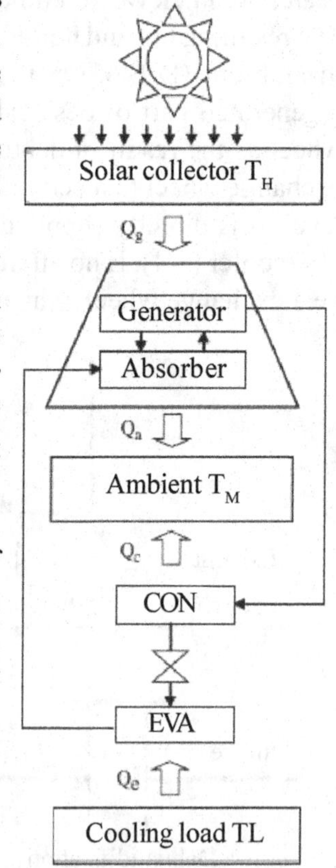

Fig. 9.7: Schematic of a closed sorption refrigeration system

Fig. 9.7 shows a schematic diagram of a closed sorption system. The component where sorption takes place is denoted as absorber and the one where desorption takes place is denoted as generator.

The generator receives heat Q_g from the solar collector to regenerate the sorbent that has absorbed the refrigerant in the absorber. The refrigerant vapour generated in this process condenses in the condenser rejecting the condensation heat Q_c to ambient. The regenerated sorbent from the generator is sent back to the absorber, where the sorbent absorbs the refrigerant vapour from the evaporator rejecting the sorption heat Q_a to ambient. In the evaporator, the liquefied refrigerant from the condenser evaporates removing the heat Q_e from the cooling load.

Comparison of solar refrigeration technologies

Although differing in technical maturity and commercial status, the various solar refrigeration technologies discussed in the previous sections are compared in terms of performance in Fig. 9.8. (Kim and Ferreira, 2008)

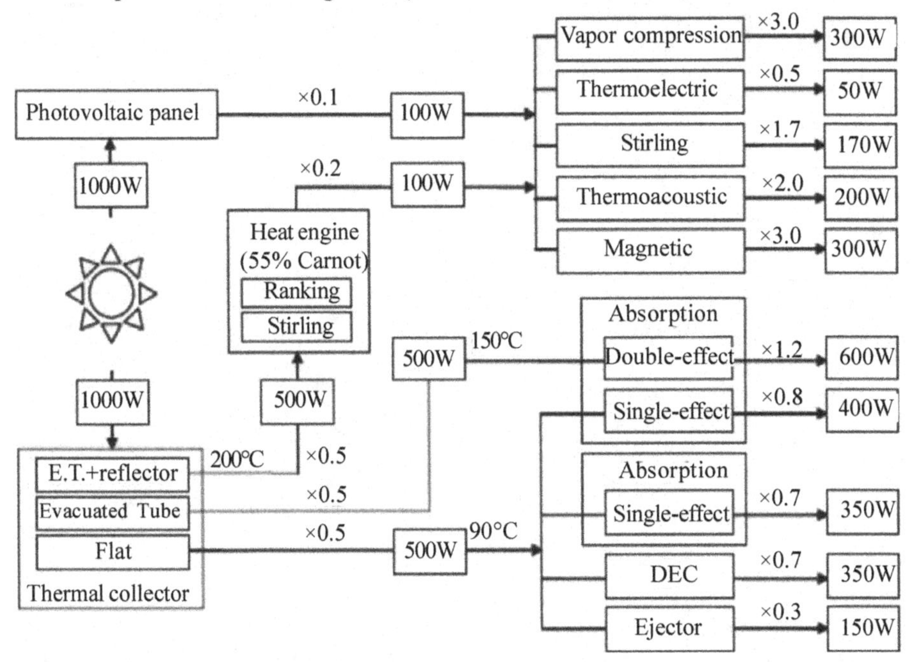

Fig. 9.8: Performance of various solar refrigeration systems

It is also noted that solar collector efficiencies listed in Fig. 8 are only indicative and will depend on ambient air temperature and solar radiation.

Summary

A variety of options are available to convert solar energy into refrigeration effect. Solar thermal with single-effect absorption system appears to be the best option closely followed by the solar thermal with single-effect adsorption system and by the solar thermal with double-effect absorption system options at the same price level. Solar thermo-mechanical or solar photovoltaic options are significantly more expensive. Desiccant systems and ejector systems will be more expensive than the first three systems.

Reference

Kim, D.S., Ferreira, C.A.T., 2009. Solar refrigeration option–a state-of-the-art review. International Journal of Refrigeration, 31,3-15.

10

Theory on Solar Air Heaters and its Application in Cold Region

R.K. Singh, Priyabrata Santra and S. Poonia

ICAR-Central Arid Zone Research Institute, Jodhpur, Rajasthan, India

Introduction

Solar air heating is the conversion of solar radiation to thermal heat. The thermal heat is absorbed and carried by air which is delivered to a living or working space. The transparent property of air means that it does not directly absorb effective amounts of solar radiation, so an intermediate process is required to make this energy transfer possible and deliver the heated air into a living space (Fig.10.1). The technologies designed to facilitate this process are known as solar air heaters. Solar air heating technologies use only free, renewable and clean energy and can help defray the rising cost of conventional energy.

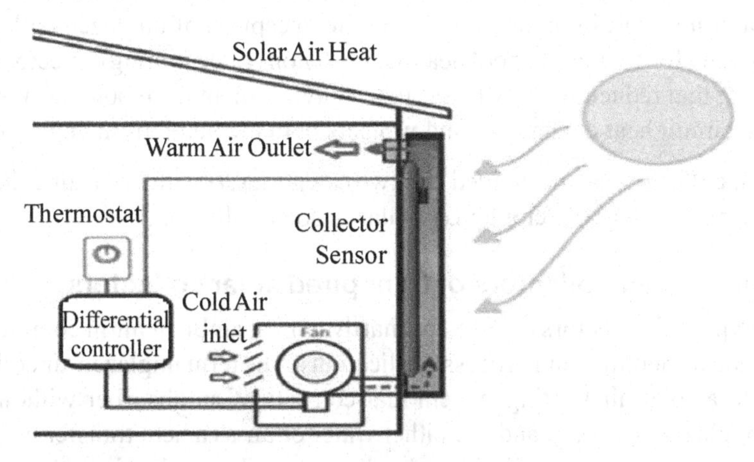

Fig. 10.1: Solar air heating system
(*Source:* MPI Solar, Bloomington, IN)

Solar air heaters operate on some of the most fundamental and simple thermodynamic principles:

- Absorption of the solar radiation by a solid body results in the body heating up. In broad terms this solid body is known as the collector. Some bodies are better at absorption than others, such as those with black non-reflective surfaces.
- Convection of heat from the heated solid body to the air as it passes over the surface. Typically a fan is used to force the air across the heated body; the fan can be solar powered or mains powered.

Solar thermal heat technologies

Different types of solar air heater technology achieve this process using the same basic principles but through the use of different solid bodies acting as the collector. The fan that transfers the air across the heated surface is also used as part of a ducting system to direct the heated air into the dwelling space. In addition to heating the air within that space, the heat can further be absorbed by thermal mass such as walls, flooring, furniture and other contents. Such heat is effectively stored and slowly dissipates beyond sunlight hours.

Non concentrating solar technologies

Non concentrating solar thermal collectors can be used for low temperature air and water heating in buildings or industry. All systems have a number of components in common, such as the absorber that collects the incoming near infrared and visible solar radiation. Most collectors have a so called selective absorber that reduces the release of infrared radiation, ensuring as much heat as possible is retained (IEA, 2012a). All collectors have a circuit through which the heat transfer fluid or air flows. With the exception of unglazed collectors, mainly used for swimming pool heating, most non concentrating collectors have a housing that reduces energy losses to the environment from both the absorber and the circuit heat exchanger, and protects both elements from degradation.

Solar air collectors can be divided into two categories: (i) Unglazed air collectors or Transpired solar collector and (ii) Glazed solar collectors

(i) Unglazed air collectors or transpired solar collectors

These types of collectors are used primarily to heat ambient air in commercial, industrial, agriculture and process applications. The term unglazed air collector refers to a solar air heating system that consists of an absorber without any glass or glazing over top and use either water or air as a heat transfer medium. The most common type of unglazed collector on the market is the transpired solar collector.

Transpired solar collectors are usually wall-mounted to capture the lower sun angle in the winter heating months as well as sun reflection off the snow and achieve their optimum performance. The exterior surface of a transpired solar collector consists of thousands of tiny micro-perforations that allow the boundary layer of heat to be captured and uniformly drawn into an air cavity behind the exterior panels. This solar heated ventilation air is drawn into the building's ventilation system from air outlets positioned along the top of the collector and the air is then distributed in the building via conventional means or using a solar ducting system.

Transpired solar collectors act as a rain screen and they also capture heat loss escaping from the building envelope which is collected in the collector air cavity and drawn back into the ventilation system. There is no maintenance required with solar air heating systems and the expected lifespan is over 30 years. Unglazed transpired collectors can also be roof-mounted for applications in which there is not a suitable south facing wall or for other architectural considerations.

(ii) Glazed air collectors

These types of air collectors are of recirculating types that are usually used for space heating. Functioning in a similar manner as a conventional forced air furnace, systems provide heat by recirculating conditioned building air through solar collectors. Through the use of an energy collecting surface to absorb the sun's thermal energy and ducting air to come in contact with it, a simple and effective collector can be made for a variety of air conditioning and process applications.

A simple solar air collector consists of an absorber material, sometimes having a selective surface, to capture radiation from the sun and transfers this thermal energy to air via conduction heat transfer. This heated air is then ducted to the building space or to the process area where the heated air is used for space heating or process heating needs.

Glazed collectors can be divided into two types:

Flat plate collectors: Flat plate collectors usually use water (including an anti freeze liquid in regions with low temperatures) as heat transfer fluid, with only a few using air. They consist of housing in the form of a shallow box, comprising an insulated casing and one or two transparent layers of low iron, tempered solar glass. Flat plate collectors can reach up to 60% efficiency and can provide heat at temperatures up to 80°C, and are thus particularly suited to provide heat in hot water systems (IEA 2011b).

Evacuated tube collectors: Evacuated tube collectors usually consist of an evacuated double glass tube which includes a metal U pipe or a heat pipe in its centre. Vacuum tubes consisting of an evacuated single glass tube with a metal absorber in its centre are also common. In both cases, the vacuum reduces heat losses to the environment and raises the heat collection efficiency. Evacuated tube collectors perform better than flat plate collectors where ambient temperatures are lower and also work better in low irradiation conditions and are thus likely to be favoured in locations where clouded skies are common. Evacuated tube collectors demonstrate high efficiency and provide heat of up to 160°C (IEA 2011b). They are therefore suitable for high demand water heating systems or for provision of process heat.

There are different types of solar thermal heating systems in buildings, the most widespread being hot water systems with a thermosiphon, that are common for instance in China, as well as systems with forced circulation that are more common in central and northern Europe. Such systems can cover up to 60% of the total energy required for hot water provision in a single family building (ESTTP 2013). Combi systems that provide both hot water and space heating are mainly used in central Europe, in particular in Germany, Austria, Switzerland and France. In well insulated buildings in central Europe, these systems provide 20% to 30% of the overall energy demand for hot water and space heating, whereas in more sunny countries up to 60% can be reached.

Concentrating solar technologies

Concentrating solar technologies focus sunlight from a large aperture area onto a small area by means of mirrors and range from simple cook stoves to high tech, large scale collectors for high temperature heat and/or power generation. The temperature range that can be achieved depends on the concentration ratio with higher concentration resulting in a higher temperature achieved. Concentrating solar technologies require clear skies and sufficient direct normal irradiance (DNI) to reach high levels of performance, which limits the areas suitable for their deployment.

The compound parabolic concentrator is a reflector type used with both non evacuated flat plate collectors and evacuated tube collectors. It is typically designed for a concentration ratio < 2 and makes use of both direct and diffuse radiation. Concentrating solar thermal technologies could also provide heat for industrial processes at the upper end of the medium temperature range (150°C to 450°C) in regions with sufficient levels of DNI.

Solar air heat applications

A variety of applications can utilize solar air heat technologies to reduce the carbon footprint from use of conventional heat sources, such as fossil fuels and to create a sustainable means to produce thermal energy. Major applications are listed below which are site specific:

(i) Industrial applications

a. Air pre-heating for combustion processes resulting into number of applications
b. Drying minerals, paper, coal, food industry products, bricks, etc.

(ii) Agricultural applications

a. Crop drying: Fruits, grains, vegetables etc.
b. Space heating for greenhouses, warehouses and animal farms etc.
c. Fruit and other produce dryers

Water and space heating alongside process heating applications are the major areas which find place in cold regions using solar thermal energy, though, it is applicable to all other areas depending on the necessity. A brief on these features is illustrated below.

Solar thermal energy use for space heating application

Space heating for residential and commercial applications can be done through the use of solar air heating panels. This configuration operates by drawing air from the building envelope or from the outdoor environment and passing it through the collector where the air warms via conduction from the absorber and is then supplied to the living or working space by either passive means or with the assistance of a fan (Chopade and Channapatanna 2016).

Solar thermal heating installations have been growing considerably around the globe in the last decade. According to Mauthner and Weiss (2012), global installed solar thermal capacity was 268 gigawatts thermal (GWth) in 2012, including district heating and industry installations. Global solar thermal energy use for heat in buildings grew at 12% per year on average, and reached 0.7 EJ in 2011. These are also used in camping purposes: (i) space heating for relief camps or military camps and (ii) space heating for camping and expeditions in cold climate

Solar thermal energy use for process heating application

Solar air heat can be used in process applications such as drying laundry, crops (*e.g.* tea, corn, coffee) and other drying applications. Air heated through a solar collector and then passed over a medium to be dried can provide an efficient means for reducing the moisture content of the material. Process heat accounts for the major share of energy consumption in industry and more than half of the heat demand, on average, is in the low and medium temperature range below 400°C. According to studies, two thirds of the medium temperature heat required in industry processes is at levels below 200°C, providing excellent opportunities for the enhanced use of solar thermal heat (IEA 2011a).

Solar thermal energy use for heat is most common in the textile industry, because solar process heat applications are most competitive within low temperature processes (e.g. textile drying in Greece). There are also a considerable number of successful examples of solar thermal energy use for heat in other industry sectors today (e.g. copper mining in Chile; metal processing/galvanization in Germany), but most of the potential is still unexploited.

Typical applications for solar heat plants are in the food and beverage industries, the textile and chemical industries and for simple cleaning processes, such as car washes. The low temperatures required in these processes (30°C to 90°C) means that flat plate collectors can be used efficiently in this temperature range. Cleaning processes are mainly applied in the food and textile industries and in the transport sector. For cleaning purposes, hot water is needed at a temperature level between 40°C and 90°C. Due to this temperature range flat plate collectors are recommended for this application. The system design is quite similar to large scale hot water systems for residential buildings, since they work in the same temperature range and the water is drained after usage. Summarizing, about 30% to 40% of the process heat demand could be covered with low to medium temperature solar collector systems.

Conclusion

Solar energy, the mother of renewable energy sources, is an inexhaustible, clean, cheap source of energy. Lying between 80 to 360 north, India has 2500 to 3200 hours of sunshine per year providing 5.4 to 5.8 kW of power per m^2 per day @ 1kJS^{-1} m^{-2}. Utilizing small portion of this immense resource would save our fossil fuels and forest without sacrificing our energy consumption. Solar hot air generation systems are more reliable, durable and cost effective energy production methods for agricultural and industry process. It is more efficient, easily adaptable from existing fuel-driven systems, environmentally friendly and hygienic.

References

Chopade, P.S. and Channapatanna, S.V. 2016. Solar air preheater performance evaluation using new design. International Journal of Engineering and Technical Research (IJETR) 4 (2): 4 pp.

ESTTP (European Technology Platform on Renewable Heating and Cooling). 2013. Strategic Research Priorities for Solar Thermal Technology, ESSTPP, Brussels.

IEA. (International Energy Agency) 2012a. Technology Roadmap: Solar Heating and Cooling, OECD/IEA, Paris.

IEA. (International Energy Agency) 2011b. Solar Energy Perspectives OECD/IEA, Paris.

Mauthner, F. and W. Weiss (2012). Solar Heat Worldwide 2012. Markets and Contribution to the Energy Supply 2010, www.iea shc.org/solar heat worldwide.

11

Animal Feed Solar Cooker, Solar Candle Device and Passive Cool Chamber: Principles and Applications

A. K. Singh and S. Poonia

ICAR - Central Arid Zone Research Institute, Jodhpur, Rajasthan, India

Introduction

The sun is an abundant source of solar energy (average value on horizontal surface 6 kW m^{-2} day^{-1}), which is freely available. This huge source of energy is non-polluting and inexhaustible in nature. It has the potential of supplementing the conventional energy sources to a great extent. There is acute shortage of conventional energy sources, which affects the overall development. Work on the utilization of such a huge, non-polluting and everlasting energy source has been carried out at CAZRI, Jodhpur for various domestic, industrial and agricultural applications in order to supplement the energy demand. This includes the development of a lot of solar thermal devices, such as, solar still, solar dryer, solar cooker, animal feed solar cooker, solar candle device and cool chamber. The details of some of these devices are given as following.

Solar cooker for animal feed

In developing countries, energy requirement for cooking purpose is generally met through firewood, which leads to deforestation. Moreover, the burning of fuelwood has adverse environmental effects since it emits large amount of CO_2 in atmosphere in the process of burning. The environmental effects of fuel wood burning have been reported in several literatures (Brunicki 2002; Tingem and Rivington 2009; Panwar et al. 2009, 2011). Keeping in mind these environmental problems of fuel wood, a transition towards low polluting energy sources for cooking purpose is required, which will also be very apt for mitigating climate change (Budzianowski 2012). Cooking with solar energy is a promising option since it is abundantly available in most parts of the world. Moreover, cooking using solar energy can be done unattended once the feed to be cooked is kept inside the cooker and thus can save considerable time which can be utilized to perform extra agricultural activity. In arid part of Rajasthan, solar irradiations are available in plenty and almost 300 clear sky days are observed. Amount of solar radiation received in the region is about 7600–8000 MJm^{-2} per annum, whereas in semi-arid region it is about 7200–7600 MJm^{-2} per annum and in hilly areas it is about 6000 MJm^{-2} per annum (Pande et al. 2009). In the arid western Rajasthan, animal husbandry contributes a major portion of the income of rural people. Livestock provides a range of benefits to rural people e.g. provides nutritious milk for domestic use, helps in income generation through sale of milk in local markets, provides manures to maintain soil fertility etc. Thus, it plays a major role in generating employment and reducing poverty in rural areas. Apart from it, livestock are commonly used for draft power in farm operations (Binswanger and Quizon 1988). However, these benefits can be availed if only digestive and nutritive feeds are given to these livestock animals. Boiling the animal feed helps in improvement of digestive and nutritional quality of the feed which in turn improve both the milk quality and quantity. Therefore, rural people in arid western Rajasthan generally boils the animal feed daily before giving it to livestocks. Firewood, cow dung cake and agricultural wastes are commonly used for boiling purpose (Nahar et al. 1996 a,b; Panwar et al. 2011). This traditional practice does not ensure the quality feed because it requires slow cooking. Solar cooking is the most suitable option to prepare the animal feed (Panwar et al. 2010, 2012). Moreover, drudgery involved in conventional boiling process can also be avoided in solar cooking and it also saves fuelwood.

The solar cookers commonly available are suitable for cooking twice a day, therefore the cost is high, whereas animal feed is to be boiled only once a day. In addition, commercially available box type cookers have low capacity and need to be oriented towards sun frequently. Therefore, it was felt that a very low cost solar cooker should be designed for boiling animal feed. Considering this, an improved solar cooker using locally available materials, e.g. clay, pearl millet husk and animal dung, has been designed, developed and tested. This new design removed the problem of orienting the cooker towards sun by providing length to width ratio of the cooker as 3:1 or more. By using the non-tracking animal feed solar cooker one can save about 30-40% of fuel requirement.

Energy balance of improved animal feed Solar Cooker

The animal feed solar cooker is based on the principle of flat plate solar collector and greenhouse effect. Shorter wave length of solar radiation enter the collector to get converted into longer wave length and get trapped inside as glass is opaque to longer wavelength. The energy balance of this solar cooker (neglecting bottom losses) was carried out by following equation:

$$pV.C_a \frac{dT_r}{dt} = \propto \tau.SA_f - UA_c (T_r - T_a) \quad (1)$$

At steady state $\rho V.C_a \frac{dT_r}{dt} \rightarrow 0$

So, $\propto \tau.SA_g - UA_c (T_r - T_a)$ (2)

or $T_r = \frac{\tau.SA_g}{UA_c} + T_a$

where U is given as, $U = \left[\frac{1}{h_i} + \frac{L}{K} + \frac{1}{h_o}\right]$

where, ρ = Density of air (kg m^{-3});

V = Volume of collector (m^{-3});

C_a = Specific heat of air (J kg^{-1} $^{\circ}C^{-1}$);

T_r = Temperature of collector (°C);

A_c = Collector area (m^2);

α = Absorptivity and transmissivity of glass of absorber;

S = Incident solar radiation (Wm^{-2});

A_f = Floor area (m^2);

U = Over all heat transfer coefficient (Wm^{-2} °C);

T_a = Ambient temperature (°C);

h_i = Inside convective radiative losses (Wm^{-2} °C)

h_o = Outside convective radiative losses (Wm^{-2} °C)

Construction and application of solar cooker for animal feed

A double glazed animal feed solar cooker with reflector was designed and fabricated at the workshop of ICAR-Central Arid Zone Research Institute, Jodhpur, India. The design was based on the concept of non tracking solar cooker wherein, length to width ratio of the cooker has been designed as more then 3:1 so that maximum amount of radiation falls on the glass window at any time in a day. The size of cooking pots was 550×450×75 mm for the boiling of animal feed. The cooker was designed for three cooking pots and size was taken accordingly. An earthen pit of 1980×760×100 mm has been dug in the ground (Fig. 11.1.). A mixture of clay, pearl millet husk and animal dung in equal proportion (volume wise) have been prepared and then water was added to make paste. The bottom of the earthen pit has been filled with the prepared paste up to a depth of 50 mm. It was left to dry in air for a couple of days. The sides of solar cooker have been made by the same paste material and then a 150 mm pearl millet husk insulation has been provided at the bottom and 24 SWG galvanised steel absorber has been put over the insulation. The absorber has been painted with black board paint. Two glass covers (4mm thick) on a removable angle iron and wooden frame have been provided over it. Three aluminium pans with lids can be kept inside the cooking chamber for boiling animal feed. The frame body of the cooker may be fabricated by an unskilled labour. Actual installation of the above animal feed solar cooker is shown in Fig. 11.2.

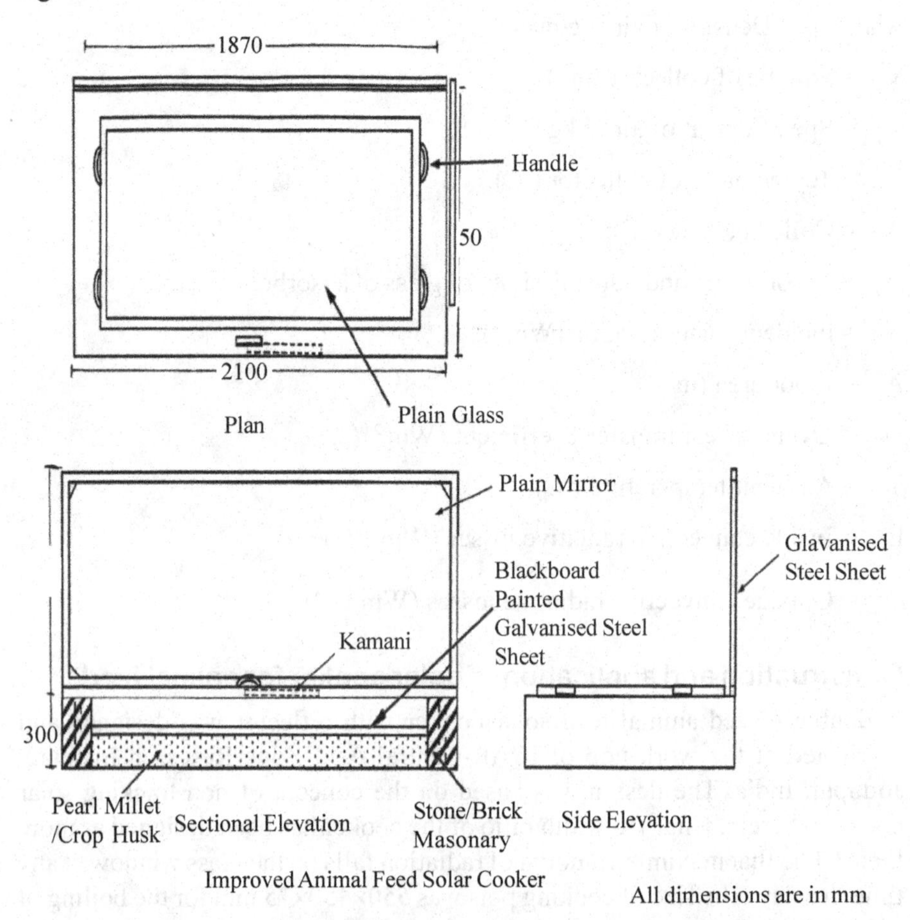

Fig. 11.1: Schematic diagram of improved animal feed solar cooker

Fig. 11.2: Improved solar cooker for animal feed installed in the field

Thermal performance and testing of solar cooker for animal feed

The thermal performance of animal feed solar cooker has been carried out according to the Bureau of Indian Standards (BIS) and American Society of Agricultural Engineers Standard (ASAE). Its first figure of merit (F_1), second figure of merit (F_2) and standardized cooking power (P_s) were found as 0.089 m^2 °C /W, 0.288 and 27.40 W, which indicate that the developed cooker falls under category "B", as per standard (F_1 >0.12, class A cooker and F_1 <0.12, class B cooker). The thermal efficiency of the animal feed solar cooker was 26.4% (Poonia et al. 2017). The maximum stagnation temperature recorded was 112°C. It can boil about 10 kg animal feed per day for milch animals and has the potential of saving about 1000 kg fuel wood annually. The cost of animal feed cooker is ₹ 9500/- without reflector and ₹ 12500/- with reflector. It has been found very suitable as the animal feed is boiled only once a day. The technology developed for the animal feed preparation not only reduces the greenhouse gas emission but also helps in fuel conservation and drudgery reduction. Meanwhile, money can also be saved, which helps to strengthen the financial status of the marginal rural farmers, if used on regular basis. Such cooking operation is done mostly by women, and they contribute significantly in the agriculture operation, thus can save time and spend more time to take care of her family or other agricultural operations.

Solar candle machine

The solar candle device is based on the principle of flat plate solar collector and greenhouse effect. The solar radiation fall on the transparent glass sheet and enter the collector and get converted into long wave thermal radiations, which is not transparent to glass surface and thus these get trapped inside and increase the inside temperature to a great extent. However, the tilt of the wax melter has to be set according to the seasonal variation of tilt angle, which is given as,

Declination angle = 23.45 [360 (284+ n) / 365]

Where n = number of day of the year, January 1, being the first day of the year.

Tilt angle = latitude ± declination angle.

The tilt remains equal to latitude (26.18 degree for Jodhpur) on March 21 and September 23. The average tilt angle for twelve months are given in Table 11.1.

Table 11.1: Average tilt angle for different months of the year

S.No	Day of month	Tilt angle
1.	January 15	48.45
2.	February15	39.80
3.	March 16	28.60
4.	April 15	16.77
5.	May 15	7.39
6.	June 14	2.87
7.	July 14	4.66
8.	August 13	10.85
9.	September 12	21.96
10.	October 12	33.90
11.	November 11	44.09
12.	December 11	48.15

The conventional methods of preparing candles from wax are unhygienic, need attendance during wax melting process and also suffer from many other drawbacks. The solar method is quite safe, convenient and obviates any type of care or attendance during intermediate melting process of raw materials. Operation and maintenance of the solar candle device is easy. The working of the device for production of candle is simple. It needs no extra space and can be operated in the house itself or in the field. One time attention is sufficient for daily production of candles/wax lamps by solar candle device. The paraffin wax is loaded once a day in the solar machine and then machine is left intact. The melting process takes place in the solar machine during the day and melted material is collected from it for making candles or wax lamps production in the evening. The time period of 2 to 3 hours in the evening is sufficient for the candle production. The candle production from a small unit of solar machine is 10-16 kg day^{-1} during summer season and 6-9 kg day^{-1} during winter season (Fig. 11.3). The dimensions of the wax melter are as given below:

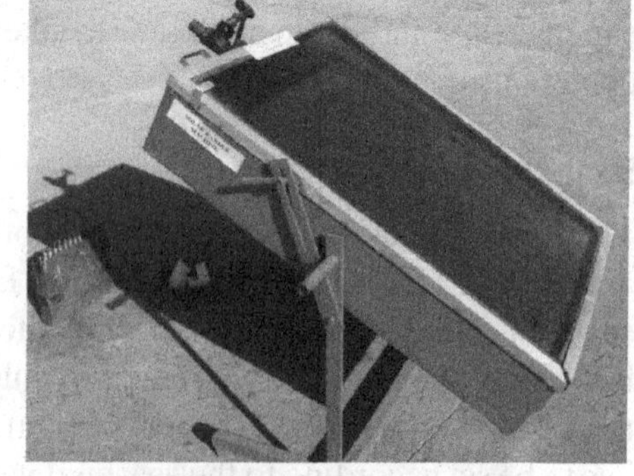

Fig. 11.3: Solar wax melting device

Absorbing area - 0.5 m^2

Loading capacity - 18kg wax

Total dimensions - 106× 75× 20 cm

The cost of the wax melter including mould was comes to about Rs.12000/-

Improved passive cool chamber

In India, the deterioration of the quality of fruits and vegetables starts immediately after harvest due to lack of farm storage. India is the second largest producer of fruit and vegetables in the world after Brazil and China. Total production of fruit and vegetables in India is about 256.10 million tonnes of which 86.60 million tonnes and 169.50 million tonnes are fruits and vegetables, respectively (Anonymous 2014). Storage of fresh horticultural produce after harvest is one of the most pressing problems of tropical countries like India. Due to high moisture content, fruits and vegetables have very short life and are liable to spoil. Moreover, transpiration, respiration and ripening processes are continued in fruits and vegetables even after harvest. Thus the deterioration rate increases due to ripening, senescence and unfavourable environmental factors. Hence, preserving fruits and vegetables in their fresh form is required to restrict chemical, bio-chemical and physiological changes to a minimum level and may be achieved through controlling temperature and humidity (Basediya et al.. 2013). Due to highly perishable nature, about 20-30% of total fruit production and 30- 35% of total vegetable production in India are wasted during various steps of the post-harvest chain (Arya et al. 2009; Kitinoja et al. 2010; Basediya et al. 2013) and the monetary losses are about Rs 2 lakh crore per annum in India (ASSOCHAM 2017).

Several simple practices are useful for cooling and enhancing storage system efficiency wherever they are used, and especially in developing countries, where energy savings may be critical. Mechanical refrigeration is, however, energy intensive, expensive, and requires uninterrupted supplies of electricity which are not always readily available. Such facilities of mechanical refrigeration system e.g. cold storages are available in India but most of them are used for storage of a single vegetable, potato. Therefore, appropriate cool storage facilities are required in India for on-farm storage of fresh horticultural produces. Low-cost, low-energy, environment-friendly cool chambers made of locally available materials, which utilize the principles of evaporative cooling, were therefore developed in response to this problem. Evaporative cooling is an environment friendly air-conditioning system that operates using induced processes of heat and mass transfer where water and air are working fluids (Camargo 2007).

Very recently Sharma and Mansuri (2017) developed solar photovoltaic (SPV) power system based evaporative cooled storage structure (ECSS) for storage of vegetables to increase shelf life. These cool chambers are able to maintain temperatures at 10–15°C below ambient, as well as at a relative humidity of 90%, depending on the season.

The evaporatively cooled storage structure has proved to be useful for short term, on-farm storage of fruits and vegetables in hot and dry regions (Chaurasia et al. 2005, Jha and Chopra 2006). Evaporative cooling is an efficient and economical means for reducing temperature and increasing the relative humidity of an enclosure, and has been extensively tried for enhancing the shelf life of horticultural produce (Jha and Chopra 2006; Okunade and Ibrahim 2011) which is essential for maintaining the freshness of the commodities (Dadhich et al. 2008).

Maintenance of low temperature is a great problem in India particularly under the hot arid conditions of Western Rajasthan commonly known as the "Thar desert" (second largest desert of the world). The hot arid zone of India, which is located in north-west India is spread in 31.7 m ha and is characterized by limited and erratic nature of rainfall, extreme temperatures with large diurnal and seasonal variation, strong solar radiation and wind regime resulting in demand for high water requirements (Rao and Roy 2012). The weather conditions, even in normal years, for most part of the year, remains too dry and inhospitable for human and livestock. Prevailing low humidity and high temperature regulates physiological activities of fresh vegetables that affects their physio-chemical characteristics during the storage period. The high ambient temperature accelerates the process of dehydration in fruit and vegetable, which leads to reduction in its water content, decrease in shelf-life and consequent spoilage in due course of time. Due to low humidity (13-33%) prevailing in the arid region particularly in summer, the cooling effect based on evaporative cooling principle becomes prominent and effective as it causes high evaporation and therefore results in more depression in temperature. Considering this, a low-cost, eco-friendly and energy saving new storage system called "Zero energy passive cool chamber (ZEPCC)" has been designed and developed at ICAR-CAZRI, Jodhpur. This system is based on evaporative cooling option for preservation and enhancing shelf - life of fruit and vegetables without using any active source of energy.

The passive cool chamber is based on the principle of evaporation. Evaporation is the process of changing liquid phase in to gaseous phase at a temperature below its boiling point. The fastest moving molecules (those with the highest kinetic energy) at the surface of the liquid have enough energy to break the attractive bonds with other molecules. They then escape the surface of the substance. Obviously, this only occurs with the molecules at the surface of the substance. Since at higher temperatures the molecules have more kinetic energy, more of them are likely to escape, and so evaporation occurs more quickly at higher temperatures. In general, evaporation occurs because systems seek equilibrium (there is a low concentration of molecules in the air, and a high concentration in the liquid). During evaporation, the required latent heat is provided by the sensible heat of adjoining air resulting in reduced air temperature. Evaporation takes place as long as there is vapour pressure deficit between wet surface and adjoining air (Psws>rh.P_{sta})

Energy balance of cool chamber: The energy balance study of improved cool chamber was carried out and the relationship for determining the temperature of cool chamber (T_{ch}) was developed as given below;

$$\rho VC_a\left(\frac{dT_{ch}}{dt}\right) = H_Q + UA_c\ (T_a - T_{ch}) - (h_p + h_{ep})\,A_c\ (T_{ch} - T_w) \tag{1}$$

At steady state condition,

$$\rho VC_a\left(\frac{dT_{ch}}{dt}\right) = 0, \text{ which leads to}$$

$$T_{ch} = \frac{\left[H_Q + UA_cT_a + (h_p + h_{ep})\,A_c\,T_w\right]}{\left[UA_c + (h_p + h_{ep})\right]} \tag{2}$$

The inside temperature of cool chamber was computed by using the developed eq (2) where H_Q = Sensible heat gain, U = overall heat transfer coefficient, A_c = surface area, T_a = ambient temperature, h_p = Convective heat loss coefficient h_{ep} = Evaporative heat transfer coefficient, T_w = wet bulb temperature, V = inside volume of cool chamber, ρ = Density of air, V = Volume of cool chamber and C_a = Specific heat of air. The depression in temperature (10-12°C) predicted was found in close proximity with the observed values for summer conditions.

Design and construction of improved zero energy passive cool chamber

The design of passive cool chamber has been improved by increasing the evaporating area and installed in the solar energy yard of CAZRI, Jodhpur (Fig. 11.4). It consists of a double walled system having inner and outer chambers made of baked bricks as shown in schematic diagram (Fig. 11.5). In both chambers bricks are stacked in vertical walls and have been joined together with cement plaster in the ratio 1:10. The inner chamber is surrounded by outer chamber and coarse sand is filled between the two. The dimensions of both chambers are 1200 mm × 1200 mm (outer chamber) and 800 mm × 800 mm (inner chamber). The heights of the chambers are 730 mm (outer chamber) and 420 mm (inner chambers). The water is also filled between inner and outer chamber. The baked bricks of cool chambers are porous enough and water filled between the cool chambers seeps through it. The water seeping through walls of outer chamber evaporates and consequently reduces the temperature of the cool chamber. The seepage of water through walls of the inner chamber also reduces temperature and as well increases humidity that provides sufficient moisture inside the chamber for preservation of vegetables under reduced temperature. The holes have been bored in both chambers by using drilling machine. In the outer chamber, 40 holes (dia 1.5cm, depth 40cm) have been bored and the distances between these holes are 12 cm. In the inner chamber, 28 holes have been bored (dia 1.5 cm, depth 20 cm) with a distance of 11.5cm between the holes. These holes have increased the evaporating area of the cool chamber for fast cooling. Provisions have also been made for water evaporation from the bottom side of the cool chamber by providing suitable

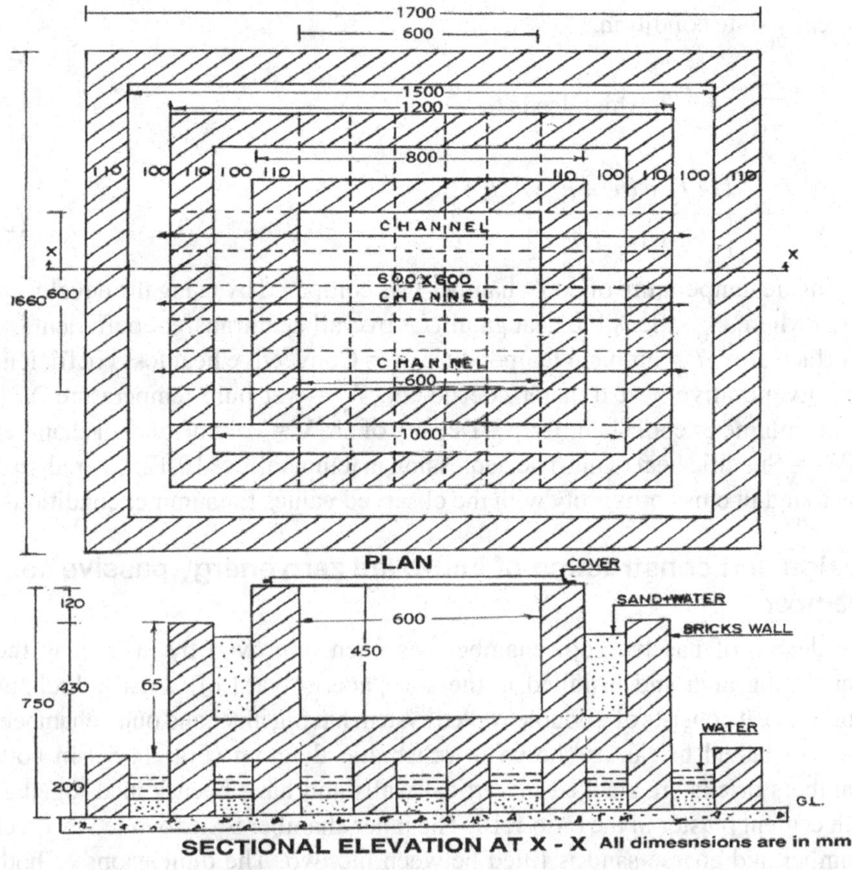

Fig. 11.4: Schematic diagram of improved low cost zero energy passive cool chamber

channels which further enhances temperature reduction and maintains high humidity in the chamber. The water filled up in the annular space of side walls helps to maintain high humidity inside the inner chamber and reduces temperature. To cut-off solar radiation, a slanting shed (3250 mm × 3000 mm) has been fabricated. The improved cool chamber was found to achieve maximum depression in temperature in one hour compared to 2-3 hours by old chamber. About 15 to 40 liters water is required daily in the cool chamber to keep the walls wet depending upon the season and the climatic conditions of the day (15 to 30 liters water in winter and 20 to 40 liters in summer). Kitchen or waste water may also be used. About 2 to 3 liters water is sprinkled on the cotton cloth provided on the top side of the lid of the cool chamber to conserve moisture that maintains high humidity inside the cooling area.

Fig. 11.5: Improved zero energy passive cool chamber for preservation of vegetables.

Performance of zero energy passive cool chamber

The improved cool chamber is able to reduce the inside temperature by about 12-14°C during summer and 6-8°C during winter and maintains humidity more than 90%, to preserve vegetables for short term period. It can safely preserve vegetables for 7 days during winter and 4-5 days during summer. The design of passive cool chamber has been improved by holes bored in outer and inner chambers. These holes have increased the evaporating area of the cool chamber for fast cooling. Provisions have also been made for water evaporation and air circulation facility from its bottom giving better results for preservation of vegetables. It successfully prolongs shelf life of vegetables and reduces weight loss, shrinkage and retains freshness of vegetables compared to vegetables preserved inside the room for a short term period. It can safely preserve vegetables for 7 days during winter and 4-5 days during summer. The improved zero energy passive cool chamber has wide utility for on-farm storage (in remote areas), vegetable markets (away from cities), retailers (vegetables vendors) and in rural areas of arid region. The cool chambers can be easily fabricated by an unskilled person with locally available materials in remote areas/villages/ rural homes as per requirement ranging from domestic use (20 kg) to commercial level (1000 kg). The above device is very useful to the farmers as well as entrepreneurs for supplementing their income. They can install these devices for enhancing shelf-life of vegetables and preserve them during transit storage for further use or onward sale. The cool chamber is recommended for preservation of vegetables for on-farm storage/vegetables markets for a period of 2 to 3 days. The vegetables stored in cool chamber, on commercial basis, for this duration remain as good as fresh and fetch good market value. For domestic purpose, the cool chamber, besides prolonging shelf life of vegetables, can also be used for preservation of left-over food materials including milk and its by-products. It can go a long way to prevent spoilage of vegetables due to lack of proper storage facilities besides saving electricity which otherwise is required for this purpose.

The above devices are very useful to the farmers as well as entrepreneurs for supplementing their income. They can install these devices for making candles, providing boiled feed to milch animals and preserve fruits and vegetables during transit storage by enhancing their shelf life.

References

Arya, M., Arya, A. and Rajput, S. P. S. 2009. An environment friendly cooling option. Journal of Environmental Research and Development 3(4): 1254-1261.

ASAE 2003. ASAE S580: Testing and Reporting of Solar Cooker Performance.

Assocham. 2017. 10th International Food Processing Summit and Awards – Food Retail, Investment, Infrastructure, New Delhi.

Basediya, A. L., Samuel, D. V. K. and Beera, V. 2013. Evaporative cooling system for storage of fruits and vegetables- A review. Journal of Food Science and Technology 50(3): 429-442.

Binswanger, H. and Quizon, J. 1988. Distributional consequences of alternative food policies in India. In: Per Pinstrup-Anderson (Ed.), Food Subsidies in Developing Countries. Johns Hopkins Press, Baltimore, MD.

Brunicki, L. Y. 2002. Sustainable energy for rural areas of the developing countries. Energy and Environment 13: 515–522.

Budzianowski, W. M. 2012. Value-added carbon management technologies for low CO_2 intensive carbon-based energy vectors. Energy 41: 280–297.

Bureau of Indian Standards 1992. BIS standards on solar cooker IS 13429: 1992, Part I, II and III, Manak Bhavan, New Delhi, India.

Bureau of Indian Standards 2000. BIS standards on solar – box type cooker - IS 13429: Part I, II and III, Manak Bhavan, New Delhi, India.

Camargo, J. R. 2007. Evaporative cooling: water for thermal comfort. An Interdisciplinary. Journal of Applied Science 3: 51–61.

Chaurasia, P. B. L., Singh, H. P. and Prasad, R. N. 2005. Passive cool chamber for preservation of fresh vegetables. Journal of Solar Energy Society of India 15(1): 47-57.

Dadhich, S. M., Dadhich, H. and Verma, R. C. 2008. Comparative study on storage of fruits and vegetables in evaporative cool chamber and in ambient. International Journal of Food Engineering 4(1): 1–11.

Jha, S. N. and Chopra, S. 2006. Selection of bricks and cooling pad for construction of evaporatively cooled storage structure. Institute of Engineers (I) (AG) 87: 25-28.

Kitinoja, L., Al Hassan, H. A., Saran, S. and Roy, S. K. 2010. Identification of appropriate postharvest technologies for improving market access and incomes for small horticultural farmers in Sub-Saharan Africa and South Asia. WFLO grant final report to the bill and Melinda gates foundation. 318 pp.

Nahar, N. M., Gupta, J. P. and Sharma, P. 1996a. Performance and testing of two models of solar cooker for animal feed. Renewable Energy 7: 47–50.

Nahar, N. M., Gupta, J. P. and Sharma, P. 1996b. A novel solar cooker for animal feed. Energy Conversion and Management 37: 77–80.

Okunade, S. O. and Ibrahim, M. H. 2011. Assessment of the evaporative cooling system (ECS) for Storage of Irish Potato, Solanum Tuberosum L. PAT 7(1): 74-83.

Pande, P. C., Nahar, N. M., Chaurasia, P.B.L., Mishra, D., Tiwari, J. C. and Kushwaha, H. L. 2009. Renewable energy spectrum in arid region. In: Trends in Arid Zone Research in India (Eds. Amal Kar; B.K. Garg; M.P. Singh and S. Kathju), CAZRI, Jodhpur. pp 210-237.

Panwar, N. L, Kaushik, S. C. and Kothari S. 2012. State of the art of solar cooking: an overview. Renewable and Sustainable Energy Review 16: 3776–3785.

Panwar, N. L., Kaushik, S. C. and Kothari, S. C. 2011. Role of renewable energy sources in environmental protection: A review. Renewable and Sustainable Energy Review 15: 1513–1524.

Panwar, N. L., Kothari, S. and Kaushik, S. C. 2010. Experimental investigation of energy and exergy efficiency of masonry-type solar cooker for animal feed. International Journal of Sustainable Energy 29: 178–184.

Panwar, N. L., Rathore, N. S. and Kurchania, A. K. 2009. Experimental investigation of open core downdraft biomass gasifier for food processing industry. Mitigation and Adaptation Strategies for Global Change 14: 547–556.

Poonia, S., Singh, A.K., Santra, P., Nahar, N.M. and Mishra, D. 2017. Thermal Performed evaluation and testing of impoved animal feed solar cooker. Journal of Agricultural Engineering 54(1): 33-43.

Rao, A. S. and Roy, M. M. 2012. Weather variability and crop production in arid Rajasthan, ICAR-Central Arid Zone Research Institute, Jodhpur, India. 70pp.

Sharma, P. K. and Mansuri, S. M. 2017. Studies on storage of fresh fruits and vegetables in solar powered evaporative cooled storage structure. Agricultural Engineering Today 41(1): 10-18.

Tingem, M. and Rivington M. 2009. Adaptation for crop agriculture to climate change in Cameroon: turning on the heat. Mitigation and Adaptation Strategies for Global Change 14: 153–168.

12

Concentrating Type of Collectors and their Potential Applications

Priyabrata Santra, S. Poonia and R.K. Singh

ICAR-Central Arid Zone Research Institute, Jodhpur, Rajasthan, India

Introduction

A solar collector, the special energy exchanger, converts solar irradiation energy either to the thermal energy of the working fluid in solar thermal applications, or to the electric energy directly in Photovoltaic (PV) applications. For solar thermal applications, solar irradiation is absorbed by a solar collector as heat which is then transferred to its working fluid (air, water or oil). The heat carried by the working fluid can be used to either provide domestic hot water/heating, or to charge a thermal energy storage tank from which the heat can be drawn for use later (at night or cloudy days). Solar collectors are usually classified into two categories according to concentration ratios: non-concentrating collectors and concentrating collectors. A non-concentrating collector has the same intercepting area as its absorbing area, whilst a sun-tracking concentrating solar collector usually has concave reflecting surfaces to intercept and focus the solar irradiation to a much smaller receiving area, resulting in an increased heat flux so that the thermodynamic cycle can achieve higher Carnot efficiency when working under higher temperatures.

Type of concentrating type solar collectors

At present, there are four available concentrated solar power (CSP) technologies: (i) parabolic trough collector (PTC), (ii) solar power tower (SPT), (iii) linear Fresnel reflector (LFR) and (iv) parabolic dish systems (PDS). Although PTC technology is the most mature CSP design, solar tower technology occupies the second place and is of increasing importance as a result of its advantages.

(i) Parabolic trough collectors

Parabolic trough collectors can concentrate sunlight with a concentration rate of around 40, depending on the trough size (Fig. 12.1). The focal line temperature can be as high as 350°C to 400°C. The key component of such collectors is a set of parabolic mirrors, each of which has the capability to reflect the sunlight that is parallel to its symmetrical axis to its common focal line. At the focal line, a black metal receiver (covered by a glass tube to reduce heat loss) is placed to absorb collected heat. Parabolic trough collectors can be orientated either in an east–west direction, tracking the sun from north to south, or a north–south direction, tracking the sun from east to west.

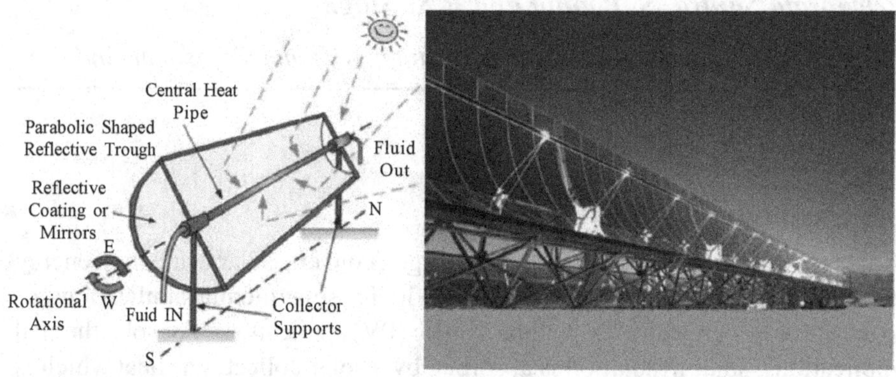

Fig. 12.1: Schematic diagram of parabolic trough collector and a CSP plant with parabolic troughs

Typically, thermal fluids are used as primary heat transfer fluid (HTF), thereafter powering a secondary steam circuit and Rankine power cycle. Other configurations use molten salts as HTF and others use a direct steam generation (DSG) system. The absorber tube (Fig. 12.2), also called heat collector element (HCE), is a metal tube and a glass envelope covering it, with either air or vacuum between these two to reduce convective heat losses and allow for thermal expansion. The metal tube is coated with a selective material that has high solar irradiation absorbance and low thermal remittance. The glass-metal seal is crucial in reducing heat losses.

Parabolic trough collectors have multiple distinctive features and advantages over other types of solar systems. Firstly, they are scalable, in that their trough mirror elements can be installed along the common focal line. Secondly, they only need two-dimensional tracking (dish-engine collectors need three-dimensional tracking, making systems more complicated), so they can achieve higher tracking accuracy than dish-engine collectors. Parabolic trough can be used for desalination, water purification and steam generation purposes.

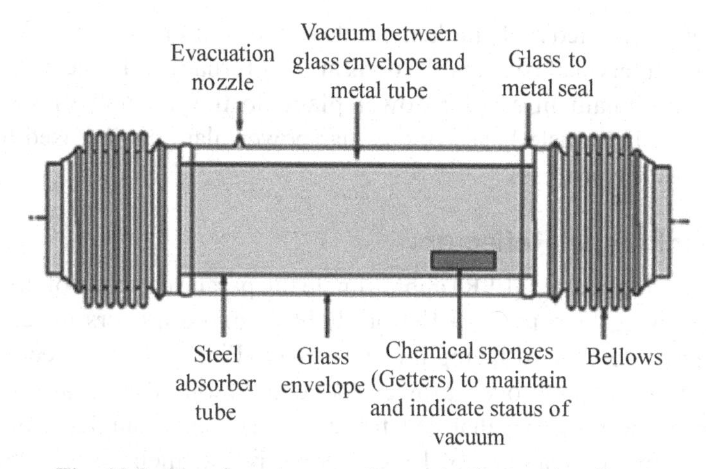

Fig. 12.2: Absorber element of a parabolic trough collector

(ii) Solar power tower

The heliostat field collector, also called the central receiver collector, consists of a number of flat mirrors/heliostats (Fig. 12.3). Due to the position change of the sun during the day, the whole array of mirrors/heliostats needs to have precise orientation to reflect incident solar lights to a common tower. The orientation of every individual heliostat is controlled by an automatic control system powered by altazimuth tracking technology. In addition, to place these heliostats with a higher overall optical efficiency, an optimised field layout design is needed.

An optimised field layout of heliostats can efficiently reflect solar light to the central tower, where a steam generator is located to absorb thermal energy and heat up water into the high-temperature and high-pressure steam (to drive turbine generators). The heat transfer fluid inside the steam generator can either be water/steam, liquid sodium, or molten salts (usually sodium nitrates or potassium nitrates), whilst the thermal storage media can be high temperature synthetic

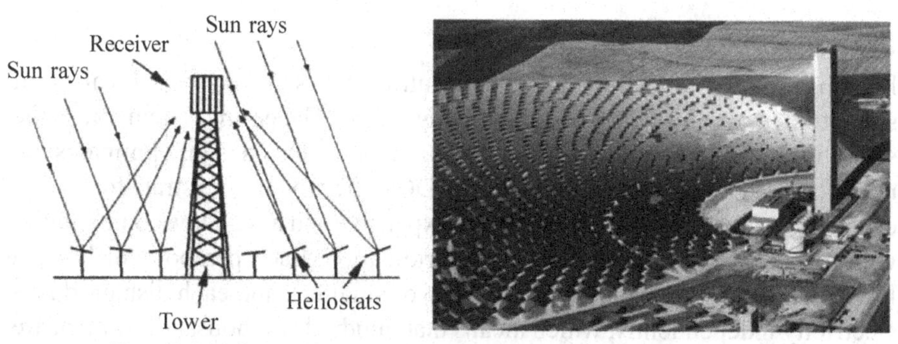

Fig. 12.3: Schematic diagram of solar power tower with heliostats

oil mixed with crushed rock, molten nitrate salt, or liquid sodium. Solar power Tower receiver has exclusive applications in solar furnace and generating power by producing steam in a solar power plant on the account of very high temperatures. The waste heat from a solar power plant can be used to run a solar desalination process.

(iii) Linear Fresnel Reflectors

Linear Fresnel reflectors (LFR) approximate the parabolic shape of the trough systems by using long rows of flat or slightly curved mirrors to reflect the sunrays onto a downward facing linear receiver (Fig. 12.4). The receiver is a fixed structure mounted over a tower above and along the linear reflectors. The reflectors are mirrors that can follow the sun on a single or dual axis regime. The main advantage of LFR systems is that their simple design of flexibly bent mirrors and fixed receivers requires lower investment costs and facilitates direct steam generation, thereby eliminating the need of heat transfer fluids and heat exchangers. LFR plants are however less efficient than PTC and SPT in converting solar energy to electricity. It is moreover more difficult to incorporate storage capacity into their design.

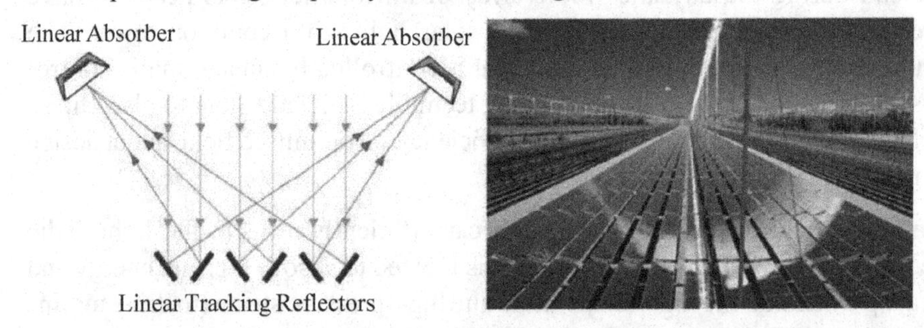

Fig. 12.4: Schematic diagram of linear Fresnel collector

Fresnel reflector can be used for various thermal applications including power generation by steam and solar chemistry systems.

(iv) Parabolic dish systems

Parabolic dish collectors (PDC), concentrate the sun rays at a focal point supported above the center of the dish (Fig. 12.5). The entire system tracks the sun, with the dish and receiver moving in tandem. This design eliminates the need for a HTF and for cooling water. PDCs offer the highest transformation efficiency of any CSP system. PDCs are expensive and have a low compatibility with respect of thermal storage and hybridization. Each parabolic dish has a low power capacity (typically tens of kW or smaller), and each dish produces electricity independently, which means that hundreds or thousands of them are required to install a large scale plant like built with other CSP technologies.

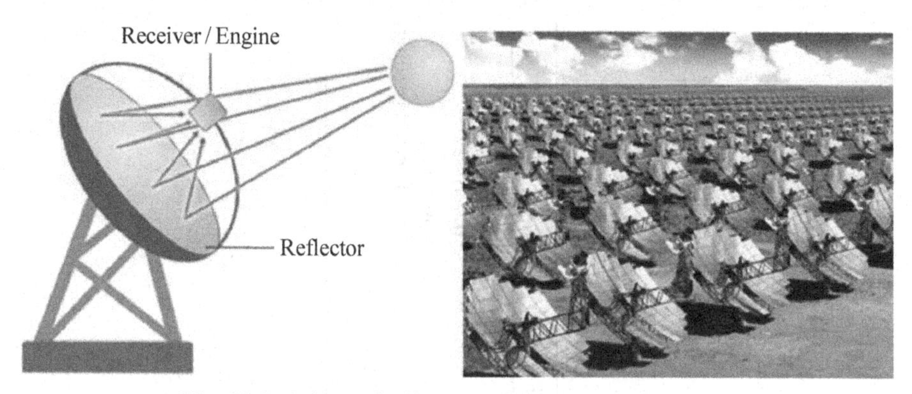

Fig. 12.5: Schematic diagram of parabolic dish system

The parabolic dish or Stirling dish has many applications from heating and cooling to underwater power systems. A Stirling engine can function in reverse as a heat pump for heating or cooling. Other uses include combined heat and power, solar power generation, Stirling cryo-coolers, heat pump, marine engines, and low temperature difference engines.

Comparison of concentrating solar collector technologies

Within the commercial CSP technologies, parabolic trough collector (PTC) plants are the most developed of all commercially operating plants. Table 12.1 compares the technologies on the basis of different parameters. In terms of cost related to plant development, solar power tower (SPT) and parabolic dish collector (PDC) systems are currently more expensive. In terms of land occupancy, considering the latest improvements in CSP technologies, SPT and linear Fresnel reflector (LFR) require less land than PTC to produce a given output. Additionally, PDC has the smallest land requirement among CSP technologies.

Water requirements are of high importance for those locations with water scarcity, e.g., in most of the deserts. As in other thermal power generation plants, CSP requires water for cooling and condensing processes, where requirements are relatively high: about 3000 l MWh^{-1} for PTC and LFR plants (similar to a nuclear reactor) compared to about 2000 l MWh^{-1} for a coal-fired power plant and only 800 l MWh^{-1} for a combined-cycle natural gas power plant. SPT plants need less water than PTC (1500 l MWh^{-1}). Dishes are cooled by the surrounding air, so they do not require cooling water. Dry cooling (with air) is an effective alternative. However, it is more costly and reduces efficiencies. The installation of hybrid wet and dry cooling systems reduce water consumption while minimizing the performance penalty.

Table 12.1: Comparison of concentrated solar collector technologies

	Parabolic troughs	Solar power tower	Linear Fresnel Reflector	Parabolic dish
Typical capacity(MW)	10-300	10-200	10-200	0.01-0.025
Operatingtemperature (oC)	350-550	250-565	390	550-750
Collectorconcentration	70-80 suns	>1 000 suns	>60 suns	>1 300 suns
Water requirement(m3/MWh)	3 (wet cooling) 0.3 (dry cooling)	2-3(wet cooling) 0.25(dry cooling)	3 (wet cooling) 0.2 (dry cooling)	0.05-0.1(mirror washing)
Annual solar to electricityefficiency(net) (%)	11-16	7-20	13	12-25

Summary

Concentrating type solar collectors are of four types: parabolic trough collector, solar power tower, linear Fresnel collector and parabolic dish. All these collectors are generally used for thermal based electricity generation by using heat provided by solar irradiation concentrated on a small area. Using mirrors, sunlight is reflected to a receiver where heat is collected by a thermal energy carrier (primary circuit), and subsequently used directly (in the case of water/steam) or via a secondary circuit to power a turbine and generate electricity. This technology is particularly promising in regions with high direct normal irradiance (DNI).

13

Solar PV Pumping System for Irrigation: Principles and Applications

Priyabrata Santra, P.C. Pande, A.K. Singh, R.K. Singh

ICAR-Central Arid Zone Research Institute, Jodhpur, Rajasthan, India

Introduction

Water is the primary source of life for mankind and one of the most basic necessities for crop production. The demand for water to irrigate the crops is increasing. For sustainable production from agricultural farms, irrigating the crops at right stages is highly important. Even in rainfed situation, lifesaving irrigation during long dry spell has also been found beneficial for crop survival and to obtain the targeted yield. Considering the depletion of groundwater below the critical zone in most part of the country, energy intensive pumping for irrigation is not a viable option. Therefore utilization of available runoff water through surface storage systems followed by pumping may be a potential solution to achieve the set goal of 'crop per drop' mission. In this connection, micro-irrigation system including drippers and sprinklers is of great importance. However, ensured power supply is essential to operate the micro-irrigation system even in remote areas. Reliable solar photovoltaic (PV) pumps are now emerging in the market and are rapidly becoming more attractive than the traditional power sources. These technologies, powered by renewable energy sources, are especially useful in remote locations where a steady fuel supply or electricity supply is problematic. About 16 million electric pumps and 7 million diesel pumps are in operations in the country for irrigation purpose; however they are highly energy intensive. Moreover, diesel operated and electrified pumps directly or indirectly emit large amount of CO_2 gas in atmosphere and hence are not environment friendly. To meet the energy demand for irrigation, solar PV pumps have been introduced under the off-grid power generation category of National Solar Mission (NSM) with a target of 1000 MW by the end of phase II (2013-2017). The target has been revised in 2015 to a total grid connected solar power generation of 1,00,000 MW comprising 40,000 MW roof top generation and 60,000 MW grid connected solar power plants (Resolution of MNRE, Govt of India, No. 30/80/2014-15/ NSM dated 1st July 2015).

Rajasthan state has already been progressed along this mission targets through installation of nearly 6000 pumps by 2013 with each pump capacity of 2200 W_p or 3000 W_p. It has been reported that Rajasthan state received the maximum share (79.36% of all India allocations) of renewable energy installations during the first phase of NSM (Pandey et al. 2012). However, long before this NSM targets, potential utility of solar PV pumps had been demonstrated through field experimentation on pomegranate orchard with drip network at ICAR-Central Arid Zone Research Institute, Jodhpur (Pande et al. 2003). Since large portion of Rajasthan state are in critical stage of groundwater utilization, further extraction of it may not be a sustainable future irrespective of the use of solar PV pumps or diesel pumps/electrified pumps. Therefore, the use of solar PV pumps for lifting water from rainwater harvesting structures or surface water reservoirs need to be promoted. In this chapter, the functionality and performance of a solar PV pumping system for irrigation purpose is discussed along with their multipurpose utility. Finally the economic analysis of solar PV pumping system is presented and compared with electrified and diesel operated pumping system.

Potential of solar PV pumping system

There exists substantial potential of using renewable sources of energy for irrigation water pumping in India (UNDP 1987; Kandpal and Garg 2003; MNES 2006). North-Wwestern India receives plentiful of solar energy. For example, average irradiance at horizontal surface at Jodhpur is 6 KWh m^{-2} day^{-1}. This energy may be utilized for operating the pump in agricultural farm for irrigation purpose. Chaurey and Kandpal (2010) also mentioned that use of renewable energy even for meeting the basic energy needs of rural communities will share a part of the huge energy demand. Recently, there is a growing interest in solar PV pump based irrigation system in the region especially after implementation of Jawaharlal Nehru National Solar Mission (JNNSM). There is an increasing need to utilize the solar PV pump based irrigation systems at different scales of operation and also for growing different crops other than orchards. Recently, several studies were carried out on the use of solar PV pump based irrigation systems and its different aspects (Derrick 1994; Hammad 1995; Suehrcke et al. 1997; Kou et al. 1998; Arab et al. 1999; Hamidat 1999; Badescu 2003; Hamidat et al. 2003; Hadj Arab et al. 2004; Manolakos et al. 2004; Hadj Arab et al. 2004; Hamidat and Benyoucef 2009;

Odeh et al. 2006; Ghoneim 2006; Glasnovic and Margeta 2007; Hamidat et al. 2007; Hamidat and Benyoucef 2008; Kaldellis et al. 2009; Meah et al. 2008; Qoaider and Steinbrecht 2010). Solar PV technology and its application in water pumping have also been reviewed by Parida et al. (2011). Financial aspect of water pumping system based on renewable energy was evaluated in detail by Barlow et al. (1993) and Purohit (2007). Meah et al. (2008) discussed some policies to make solar photovoltaic water pumping (SPVWP) system as the appropriate technology for several arid regions. Bakelli et al. (2011) optimized different components of photovoltaic water pumping system (PWPS) using water tank storage.

Components of solar PV pumping system

A solar PV system mainly comprises of i) PV panels (ii) mounting structure (iii) pump unit (AC/DC) and (iv) tracking system (Fig. 13.1).

Sizing of PV panel depends on the capacity of pump to draw water. If the suction head is about 4-5 m, which is applicable in case of a surface water reservoir, 1 hp capacity pump is sufficient which requires about 900 W_p panel in case of DC pump and 1400 W_p panel in case of AC surface pump. If the solar

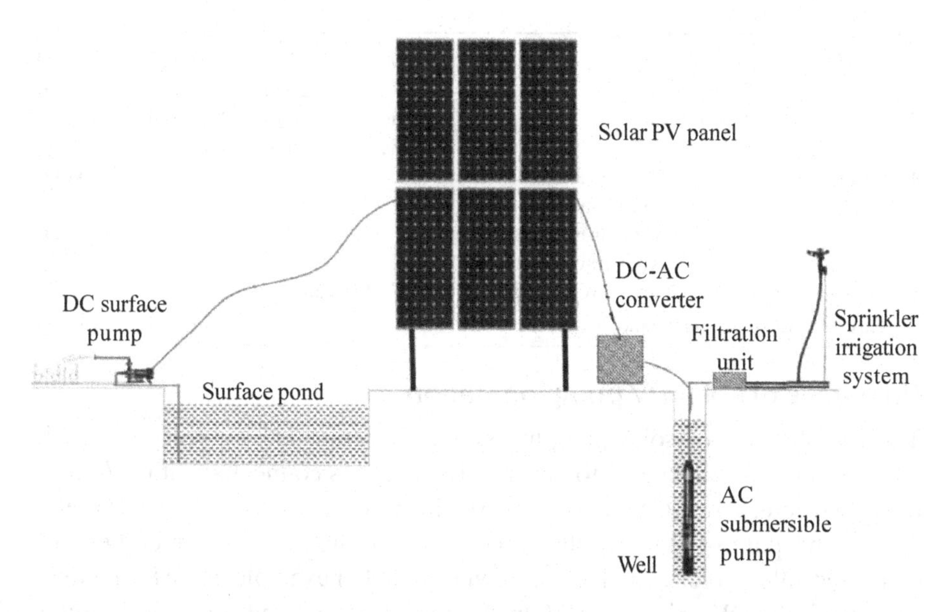

Fig. 13.1: Schematic diagram of a solar PV pumping system

PV pump is to be used for drawing more deep water from wells or tube wells, panel size will be higher accordingly. The mounting structure for erecting the panels with an angle from horizontal surface, which is generally equal to the latitude of any place needs to be strong enough to withstand the wind forces. The pumps to be used in a solar pumping system may be either DC or AC type and surface or submersible type as per situation. As the PV panels generate DC current, additional DC-AC inverter system is required for AC pumping system. To track the panel perpendicular to the sun, tracking system is required. Two types of tracking system are available i) one axis tracking which tracks the solar panel as per azimuthal rotation of sun from east to west, ii) in additional to azimuthal rotation PV panels can be tracked as per zenith angle of sun using a two axis tracking system. Both manual and auto tracking systems are available in the market. However, in case of auto tracking system there will be an additional cost of tracker. Cost of available solar PV pumping system in market with 3 hp capacity pump is about Rs 4.00 lakhs with additional cost of Rs 14,000/- for auto tracker and about Rs 8,000/- for providing lighting systems (Table 13.2).

Table 13.2: Approximate base price of solar PV pumping system in Rajasthan state during 2015

Details	DC/AC	Mounting structure	Head (metre)	Base rate (Rs)	
				3 HP	5HP
SPV surface pump	DC	Static	20 m	3,87,887	6,20,000
	AC	Static	20 m	4,10,000	5,70,000
SPV submersible pump	DC	Static	20 m	4,49,513	6,21,068
	AC	Static	20 m	4,14,578	5,80,000
Additional cost	50 m head over 20 m				6,500
	75 m head over 20 m				11,000
	Manual tracking system				4,050
	Auto tracking system				17,500
SPV domestic lighting system 37 W_p/40 Ah battery/9 W×2 fixtures					7,999
Fencing around solar panels and structure					14,000

Principles of solar PV pump operation

The PV modules of solar pumping system generate DC current as soon it receives solar irradiance on top of it. A DC pump is connected with PV array directly whereas for AC pump an DC-AC inverter is required in between them. As per the pump type and characteristics, the arrays of PV modules are combined either in series and parallel connection. For example, if the PV module capacity is 200 W_p with 53 V_{oc} (Open circuit voltage) and 5 I_{sc} (short circuit current), the modules are connected in series to operate 1 HP AC pump because it requires about 220-240 V AC after conversion of DC to AC by the inverter. In case of DC pump with characteristics voltage and ampere rating of 65 V and 12 A, the modules are connected in parallel to obtain the desired ampere to operate the pump at its full potential. Once, sufficient voltage and ampere is supplied to pump, it withdraws water from surface water reservoir or wells. Deeper is the groundwater or suction head in surface water reservoir, larger capacity of pump is required to withdraw water, which increases the PV panel size. Once PV operates with the electricity generated by PV modules, next part is to design the irrigation system as per the PV panel size and pump capacity. If an irrigation network which requires higher operating pressure than the pump creates as per its capacity and available solar irradiance, the pump will not able to irrigate the crops. For example, 1 HP solar pump can generate about 1.5-2 kg cm^{-2} operating pressure at full radiation during day time and hence, sprinklers of 2 kg cm^{-2} cannot be connected with it. For this purpose, larger size PV pumping system e.g. 3 HP or 5 HP solar PV pumping system is required.

Application of solar PV pumps

Solar PV pumps can be best used with pressurized irrigation system e.g. drippers, sprinkler etc. Small sized solar PV pumps of 1 HP capacity is best suitable to irrigate crops from surface water reservoir in to greenhouses, poly houses, shed net houses for high-value vegetable production. Larger size solar pumps (e.g. 3 HP and 5 HP capacity) can be used in canal command areas to irrigate crops with sprinklers. Even, a 5 HP pumping system can withdraw groundwater from 75 m below ground level. However, considering the critical situation of groundwater depths at many places of India, connection of a solar pump with groundwater may be avoided. Solar PV pump has a huge potential for irrigating pomegranate orchard through drippers. Even, farmers are extensively using solar PV pumping system for growing mustard, wheat, groundnut etc. in Indira Gandhi Nahar Pariyojana (IGNP) command areas of Ganganagar and Hanumangar district.

Conclusion

Solar PV pumping systems has been viewed as one of the most viable options for future energy secured agriculture and a significant progress has been made in states like Rajasthan and Gujarat. Analysis of pump performance as per availability of solar radiation, it has been observed that the solar PV pumps can be operated easily for 6 hours a day from morning 10:00 am to afternoon 4:00 pm. Moreover, 1 HP solar pump system was found satisfactory to operate different efficient pressurized irrigation systems and even mini-sprinklers, which require an operating pressure of about 2 kg cm^{-2}. Protected agriculture system requiring to run a fan motor was also successfully operated by changeover switching facility of solar pumps. Even a portion of domestic electricity need of farmers can be fulfilled by solar PV pumping system. Comparative analysis of solar PV pumps with diesel operated pumps and electrified pumps revealed that solar pumps will be highly beneficial to farmers.

References

Arab, A.H., Chenlo, F., Mukadem, K., Balenzategui, J.L., 1999. Performance of PV water systems. Renewable Energy 18, 191–204.

Badescu, V., 2003. Time dependent model of a complex PV water pumping system. Renewable Energy 28 (4), 543–560.

Bakelli, Y., Hadj Arab, A., Azoui, B., 2011. Optimal sizing of photovoltaic pumping system with water tank storage using LPSP concept. Solar Energy 85 (2011) 288–294.

Barlow, R., McNeils, B., Derrick, A., 1993. Solar pumping: An introduction and update on the technology, performance, costs and economics. World bank technical paper no 168. Intermediate Technology Publications and the World Bank, Washington DC, pp. 153.

Chaurey, A., Kandpal, T.C., 2010. Assessment and evaluation of PV based decentralized rural electrification: An overview. Renewable and Sustainable Energy Reviews, 14(8), 2266-2278.

Derrick, A., 1994. Solar photovoltaics for development: progress and prospects. Renewable Energy 5 (1–4), 229–236.

Ghoneim, A., 2006. Design optimization of photovoltaic powered water pumping systems. Energy Conversion and Management 47 (11–12), 1449–1463

Glasnovic, Z., Margeta, J., 2007. A model for optimal sizing of photovoltaic irrigation water pumping systems. Solar Energy 81 (7), 904–916.

Hadj Arab, A., Benghanem, M., Chenlo, F., 2006. Motor-pump system modelization. Renewable Energy 31 (7), 905–913.

Hadj Arab, A., Chenlo, F., Benghanem, M., 2004. Loss-of-load probability of photovoltaic water pumping systems. Solar Energy 76 (6), 713–723.

Hamidat, A., 1999. Simulation of the performance and cost calculations of the surface pump. Renewable Energy 18, 383–392.

Hamidat, A., Benyoucef, B., 2008. Mathematic models of photovoltaic motor-pump systems. Renewable Energy 33, 933–942.

Hamidat, A., Benyoucef, B., 2009. Systematic procedures for sizing photovoltaic pumping system, using water tank storage. Energy Policy 37, 1489–1501.

Hamidat, A., Benyoucef, B., Boukadoum, M., 2007. New Approach to Determine the Performances of the Photovoltaic Pumping System. Revue des Energies Renouvelables ICRESD-07 Tlemcen, pp. 101–107.

Hamidat, A., Benyoucef, B., Hartani, T., 2003. Small-scale irrigation with photovoltaic water pumping system in Sahara regions. Renewable Energy 28 (7), 1081–1096.

Hammad, M., 1995. Photovoltaic, wind and diesel: a cost comparative study of water pumping options in Jordan. Energy Policy 23 (8), 723–726.

Kaldellis, J. et al., 2009. Experimental validation of autonomous PV-based water pumping system optimum sizing. Renewable Energy 34 (4), 1106–1113.

Kandpal, T.C., Garg, H.P., 2003. Financial Evaluation of Renewable Energy Technologies. Macmillan India Ltd., New Delhi, India.

Kou, Q., Klein, S.A., Beckman, W.A., 1998. A method for estimating the long-term performance of direct-coupled PV pumping systems. Solar Energy 64 (1–3), 33–40.

Mani, A. 1981. Handbook of solar radiation data for India. Allied Publishers Private Limited, DST, New Delhi.

Manolakos, D., Papadakis, G., Papantonis, D., Kyritsis, S., 2004. A stand-alone photovoltaic power system for remote villages using pumped water energy storage. Energy 29, 57–69.

Meah, K., Fletcher, S., Ula, S., 2008. Solar photovoltaic water pumping for remote locations. Renewable and Sustainable Energy Reviews 12 (2), 472–487.

MNES, 2006. Annual Report: 2005–06. Ministry of Non-Conventional Energy Sources (MNES), New Delhi.

Odeh, I., Yohanis, Y., Norton, B., 2006. Economic viability of photovoltaic water pumping systems. Solar Energy 80 (7), 850–860.

Pande, P.C. et al., 2003. Design development and testing of a solar PV pump based drip system for orchards. Renewable Energy 28 (3), 385–396.

Pandey, S., Singh· V.S., Gangwar, N.P., Vijayvergia, M.M., Prakash, C., Pandey, D.N., 2012. Determinants of success for promoting solar energy in Rajasthan, India. Renewable and Sustainable Energy Reviews, 16(6), 3593-3598.

Parida, B., Iniyan, S., Goic, R., 2011. A review of solar photovoltaic technologies. Renewable and Sustainable Energy Reviews 15 (2011) 1625–1636.

Purohit, P. 2007. Financial evaluation of renewable energy technologies for irrigation water pumping in India. Energy Policy 35, 3134–3144.

Qoaider, L., Steinbrecht, D., 2010. Photovoltaic systems: a cost competitive option to supply energy to off-grid agricultural communities in arid regions. Applied Energy 87 (2), 427–435.

Suehrcke, H., Appelbaum, J., Reshef, B., 1997. Modelling a permanent magnet DC motor centrifugal pump assembly in a PV energy system. Solar Energy 59 (1–3), 37–42.

UNDP, 1987. Global Windpump Evaluation Programme: Country Study on India-Preparatory Phase. World Bank and United Nations Development Programme, Amersfoort.

14

PV–Hybrid Structures for Cultivation of Crops in Arid Region

A. K. Singh, S. Poonia and Priyabrata Santra

ICAR - Central Arid Zone Research Institute, Jodhpur, Rajasthan, India

PV clad structures and hybrid devices for rural and agricultural applications have been developed after considering different design aspects, climatic parameters, solar radiation availability on different planes, heat transfer coefficients, performance of PV in arid region and inter phasing of different components while keeping in view, practicality, application and ease of operation. Performance of the developed systems was studied, mathematical model developed and energy savings worked out for techno-economic considerations. The chronology and details of the developed PV structures and hybrid devices are as follows:

Green structure for environmental control

First of all a Quonset type environment control structure having a length of 5.3 m and width 4 m was fabricated with angle iron and iron rods and it was covered by agro-net (75%) having surface area of about 55 m^2. The structure having a volume of 33 m^3 was fixed with long side along east west direction and had provisions for incorporation of misting unit on west side and a door on east.

The cooling system comprises AC operated mister, which is primarily a fast moving disc pivoted at the axle of an AC motor (50W) for generating mist from water, which is lifted and circulated around the disc by a small submersible pump (18 W AC), fixed in a steel water tank. In order to create cooling effect inside the enclosure, an arrangement was made to pass a fast moving stream of air using a DC fan assembly (40 W). The fan was fixed on an especially designed chamber with provision to regulate speed and direction of air towards created fine mist. A PV system was developed that comprises PV panel (70 Wp), storage battery and inverter with provision for operation of both AC and DC loads and the panel holder was provided with wheels for ease in mobility. The complete system (Fig. 14.1) was operated to ascertain its functionality.

Fig. 14.1: Green structure with PV mister

Energy balance of green structure

Energy balance study of environment control enclosure was carried and the inside temperature (T_i) was calculated by using the developed equation $T_i = T_a + (0.12H\,Ag)/(0.33NV + UAc)$

Where

U = Overall heat transfer coefficient (°C)

Ti = Inside temperature (°C)

T_a = Ambient temperature (°C)

Ac = Surface area (m^2)

N = Number of air change per hour

V = Inside volume of structure

Ag = Ground area (m^2) and

H = Solar radiation (Wm^{-2})

Assuming N as 20, H is 800 Wm^{-2} and T_a is 20°C, the maximum temperature rise was calculated to 4.1°C above ambient temperature.

Tomato seedlings were planted inside the structure and T_i was recorded about 4°C higher than T_a during winter month, which is in close proximity with the theoretically computed values. During summer months reduction in temperature by 2.5-3°C was observed near the crop with the use of mister. Energy balance components inside the enclosure were recorded and a steady state mathematical model was developed. The inside observed data were in close proximity with the theoretically evaluated values using the developed model.

During March-April, tomato yield was monitored and found on an average 500 gm^{-2}, approximately 50% higher than the yield obtained from plants grown without enclosure. The growth of the crop inside and outside can be seen in Fig. 14.2, which clearly indicates the advantage of the green structure.

Subsequently, the mister, operated by PV assembly with provision for operation of both AC and DC loads provided a cooling of 2.5-3°C, again in close proximity to theoretically evaluated values. However, in extreme summer the cooling system requires further improvisation and therefore the PV clad structure was developed.

(a) (b)

Fig. 14.2: Tomato plants growth (a) inside green structure and (b) outside in open

PV clad structure

The PV based structure was designed considering the solar radiation on vertical and inclined planes in different seasons, use of uniform temperature underneath the earth, regulation of the temperature inside the structure, in situ fixing of PV array on the roof for providing energy while creating shade and other practical considerations. In this connection the performance of amorphous silicon modules on different base materials was also considered.

The PV clad structure (ground area 15.3 m^2) was erected on iron angle frame covered with fibre sheets all around (Fig. 14.3). A door on the east side and two interconnected PV arrays of amorphous silicon solar cells (60 W_p each) were provided at the top on the slots of inclined green corrugated fibre glass sheet roof. Ten PVC pipes (7.6 cm diameter) were provided at the bottom of front and rear side and openings at the top of the sides were created with wire mesh covers for facilitating natural circulation of air. Considering the limitation of natural circulation to regulate the ambience inside the structure, preliminary observations studies on PV driven earth tube system were initiated.

Fig. 14.3: PV clad structure for regulating ambience

Earth tube cooling/heating system

Soil moisture up to 90 cm soil depth under bare soil surface as well as under cover of albedo modifying material was monitored. Soils at 90 cm remained wet in comparison to surface, although, soil moisture under cover of albedo modifying materials was found slightly higher than bare soil (Fig. 4). Since the temperature of earth does not vary much at such a depth, the possibility of better regulation of air temperature passing through pipes laid at 120 cm soil depth from surface was explored for PV-structure.

Embedded earth heat exchange pipes

Four pipes, two of RCC and two GS, (each of 5.4 meter length and 20 cm diameter) were laid at about 120 cm below the surface in a trench dug on the west side of the structure. The pipes were interconnected at the bottom and then embedded in soil. Two stone structures were prepared, one at the inlet and other at the outlet of the pipe network to regulate and monitor air flow. A small PV driven DC fan was fixed at the outlet on an iron angle frame to provide suction of air through the pipes. The outlet was provided with a shield cum regulator to direct the stream of the air towards desirable direction.

A reduction in air temperature at the outlet was observed with the operation of the PV fan when ambient temperature was above 40 ^{0}C, whereas a rise was recorded when ambient temperature was less than 20 ^{0}C, indicating the system was facilitating heat transfer in between the flowing air and the embedded pipes. PV mister was incorporated to provide more cooling in extreme summer.

Performance of PV mister based enclosure

The performance of PV based controlled environment enclosure was evaluated under ventilation and cooling modes. With PV operated mister (50 W) and fan the temperature could be reduced to 2.5–4.5 °C below ambient temperature. Energy balance components inside the enclosure were recorded and steady state mathematical model was developed for mister based cooling. The inside temperature can be calculated using the following equation;

$$T_{i=} Ta - \left[\frac{\eta_c mL}{3600} - (\alpha\tau) HA_g\right] / UA_c$$

where

U = Overall heat transfer coefficient

T_i = Enclosure temperature (°C)

T_a = Ambient temperature(°C)

A_c = Surface area (m^2)

A_g = Ground area(m^2)

H = Solar radiation ($W.m^{-2}$)

m = Amount of water evaporated per hour (kg)

η_c = Efficiency of evaporation (%)

L = Latent heat of vaporization (kJ/kg)

$\alpha\tau$ = Absorption transmissivity product

The observed data were found in close proximity to values predicted by the developed model except in very hot conditions. Certain modifications were incorporated to improve its performance in extreme weather conditions.

Agro net (75%) was provided at the top of the rear side and more shade was created at the west and the front side (Fig. 14.5a) to reduce the inside temperature during extreme summer. Guides were provided for uniform distribution of mist air (Fig. 14.5b). The observed data during different seasons indicated reduction in air temperature at the exit of the fan to a range of 6-10°C in summer and 2-4°C rise in winter, close to predicted values through the developed mathematical model. However, the reduction in the enclosure temperature compared to ambient during summer was only 4-5°C.

Performance of the PV clad enclosure (15.3 m^2) was studied with tomato crop grown in 6 m^2 area. The temperature rise inside the enclosure was more than 2-3°C during winter, and could be maintained to the limits of ambient temperature even in extreme summer with the crop stand. Approximately, 30 kg of tomatoes were produced in 12 harvesting. These results indicated regulation of the enclosure's ambience for better growth of the plants (Fig. 14.4b) by thermally modulated air after passing through earth embedded pipes.

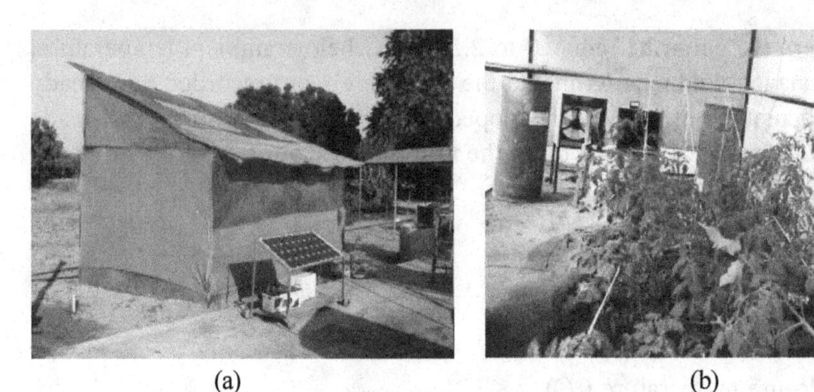

(a) (b)

Fig. 14.4: Modified PV clad structure

Thermal modelling of PV clad enclosure coupled with combined earth tube and mister

A thermal model with the combined effect of underground earth pipe and mister was developed for estimating enclosure temperature (T_r)

$$T_r = \left[\alpha\tau SA_g \frac{\beta}{\beta+1} + UA_c T_a + m_a C_a \left(T_w + T_l\right) \right] / \left(m_a C_a + UA_c\right)$$

where, T_r = Enclosure temperature (°C)

a = Absorptivity

t = Transmittance

S = Insolation (W m^{-2})

A_g = Ground area (m^2)

b = Bowen ratio

U = Overall heat loss coefficient

A_c = Surface area of enclosure (m^2)

T_a = Ambient temperature, (°C)

m_a = Air mass flow rate (kg s^{-1})

c_a = Specific heat of air (J kg^{-1})

T_w = Wet bulb temperature (°C)

T_1 = Temperature at the exit of earth pipe (°C).

The calculated values of temperature inside enclosure were found to be in close proximity with the observed values for both tomato and chilli crops grown in separate experiments (Fig 14.5). Further, the addition of PCM based storage and selected shading arrangement with this combined cooling system has been found suitable during summer.

Fig. 14.5: Tomato and chilli crops inside PV clad structure

Energy pay back and techno-economics

Energy pay back analysis of the PV clad structure was carried out. Considering the energy used for operating the suction fan throughout the year and misting unit during summer and autumn months, about 200 kWh energy could be saved through the use of photovoltaic cell. The system costs about Rs. 33,000. Although PV module will last for more than 20 years, the system's life has been taken as 10 years. Assuming a benefit of Rs. 8000 a year for nursery and growing vegetables and other crops, the benefit cost ratio was worked out to be 1.5.

Epilogue

The huge land mass of earth at about 5-10 feet depth works as source during winter and as sink during summer. The temperature at this depth varied between 22 to 28° C throughout the year. It can be used to moderate the inside temperature of a greenhouse. The addition of PCM based storage and selected shading arrangement with this combined cooling system (earth tube heat exchanger and misting) has been found suitable during summer.

15

Solar PV Devices for Application of Chemicals in Crop Fields: PV Operated Duster and Sprayer

Priyabrata Santra, P.C. Pande, R.K. Singh, D. Jain, S. Ansari, S. Thakur, P.C. Bawankar

ICAR-Central Arid Zone Research Institute, Jodhpur, Rajasthan, India

Introduction

Agriculture has been the back bone of Indian economy and culture and it will be continued to remain as such for a long time in future. Parallel to this, energy security of a country is also very important and efforts have been given on renewable energy utilization since the fossil fuel based energy is depleting at a very fast rate. In agricultural fields, considerable amount of energy is used to do different field activities e.g. ploughing, irrigation through pumps, intercultural operations, spraying of agricultural chemicals for plant protection, harvesting, post-harvest processing etc. Therefore, there is also need to replace the conventional energy source with renewable sources to operate above mentioned agricultural activities.

Approximately, 35% of the crop production is damaged if pest and diseases are not controlled at right time. Uniform spraying of liquid formulations throughout the crop field is very important for effective control of pest and diseases. Using sprayer, liquid pesticide formulations are generally broken down to minute droplets of effective size for uniform distribution over a large surface area. Different types of sprayers are used in agricultural field based on different requirements.

Ultra-low volume spraying	: $<5\ l\ h^{-1}$
Low volume spraying	: $50\text{-}150\ l\ h^{-1}$
High volume spraying	: $250\text{-}500\ l\ h^{-1}$

On the basis of energy employed to atomise and eject the spray fluid the sprayers are categorized as: (i) hydraulic energy sprayer, (ii) gaseous energy sprayer, (iii) centrifugal energy sprayer and (iv) kinetic energy sprayer. Dose of agricultural chemicals also plays a critical role since under dose may not give the desired coverage whereas overdose is expensive and may contaminate the food chain through residues. Apart from spraying dusting of agricultural chemicals e.g. sulphur dust, malathion dust are often recommended to control pest and diseases in crop field. Therefore, design and development of sprayer and duster is essential for different type of field and crop conditions.

Solar PV duster

The solar PV duster (Fig. 15.1) essentially comprises a photovoltaic panel carrier, storage battery and especially designed compatible dusting unit. The PV panel is carried over the head with the help of a light PV panel carrier, which provides shade to the worker and simultaneously charges the battery to run the duster. The battery is stacked in a bracket, which is fixed in situ to the panel carrier. Approximate cost of this device is about ₹ 9000/- and for dissemination of it, MoU has been signed between ICAR-CAZRI, Jodhpur and a private manufacturing unit. The unit has also the additional facility for lighting purpose during night time. The unit has been successfully tested for dusting sulphur dust and malathion powder.

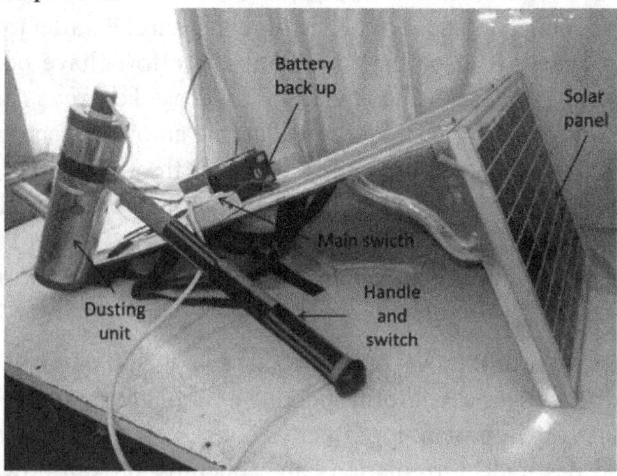

Fig. 15.1: Solar PV duster

Solar PV sprayer

To provide energy to DC pump (60 W) of the PV sprayer (Fig. 15.2), 120 W_P capacity (60 W_P each) solar PV modules are connected so that the produced energy may be directly used by DC motor. To provide continuous supply of power to the system and its other uses, a provision of battery bank (two batteries 12V, 7Ah each) was made.

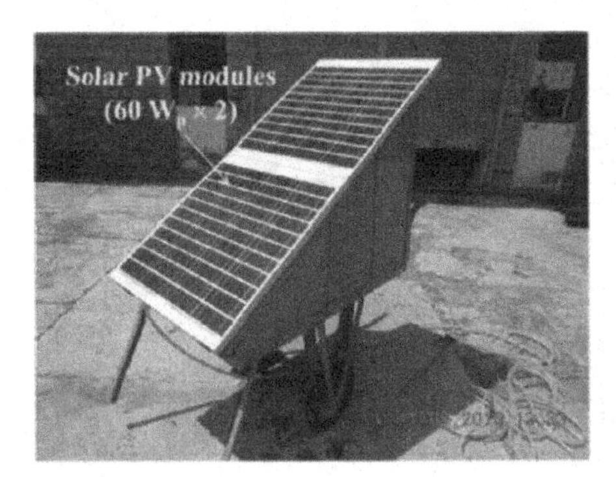

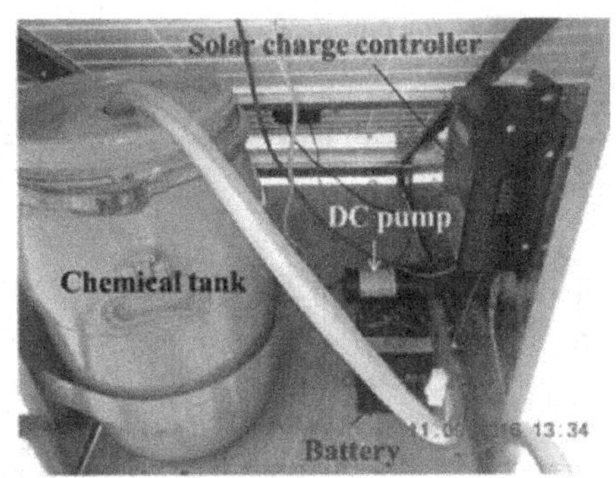

Fig. 15.2: Solar PV sprayer

The PV sprayer was designed with following units: (i) energy conversion unit for generating electricity from solar irradiation using solar PV module, (ii) energy storage unit in the form of battery, (iii) DC motor with pumping system and (iv) sprayer unit. The specification of PV sprayer is listed in below Table 15.1.

Table 15.1: Technical specifications of different components of the sprayer

Sr No.	Components	Specification
1.	DC pump (Diaphragm pump)	Diaphragm pump power: 60W Volts: 24V Amps: 2.5A Suction height: 2 m Working pressure: 82 PSI Open flow: 5.0 LPM
2.	Solar PV modules of 60 W_p capacity (2′60 W_p)	Power, W_p: 120 W_p, Module efficiency: 13.1 %
3.	Battery (2 Nos)	Operating voltage: 12V Current: 7 Ah Output power: 24 V, 7 Ah
4.	Charge controller	24V, 20A
5.	Nozzle (Hollow cone)	Discharge rate: 900 cm^3 min^{-1}; Spray angle: 60°

Cost of PV sprayer

The detailed cost of different components of the developed sprayer is presented in Table 15.2. The system is designed in such a way that its use will be multi-purpose. It provides additional facility to farmers for fulfilment of domestic use of energy.

Table 15.2: Cost of different components of developed solar PV sprayer

Sr. No.	Component	Cost (₹)
1.	Solar panels	6000
2.	DC motor	5500
3.	Batteries	1500
4.	Charge controller	1200
5.	Trolley	9500
6.	Pesticide tank	200
6.	Accessories (Nozzle, pipe, plywood, GI sheet, paints etc.)	3500
	Total	27400

Summary

Solar PV duster and solar PV sprayer was developed for dusting and spraying of agricultural chemicals in agricultural field for protection from pests and diseases. Apart from plant protection, PV sprayer can be used for spraying water on crops for reducing the heat stress. In both the devices, options are kept for using it as lighting source through LED during night time. Therefore, both the devices may be quite useful to farmers.

16

Field Performance of a Solar PV Pumping System: A Case Study with 1 HP Solar Pump

Priyabrata Santra, A.K. Singh, P.C. Pande, and R.K. Singh

ICAR-Central Arid Zone Research Institute, Jodhpur, Rajasthan, India

Introduction

For optimum use of harvested rain water in surface reservoir like farm ponds or tankas, small sized solar PV pumping systems with 1 hp AC and DC motor were experimentally tested at research farms of ICAR-Central Arid Zone Research Institute, Jodhpur (Fig. 16.1). Total suction head in both pumps was about 5 m. Among two installed pumps, the system with AC pump consisted of 1400 W_p (200 $W_p \times 7$) whereas the DC pump consisted of PV array of 920 W_p (230$W_p \times 4$). Each PV panel of AC solar pump was connected in series, which was further connected with an inverter to generate AC output of about 220-240 volt and 4-4.2 ampere. In case of DC solar pump, panels are connected in parallel to generate DC output of 40-50 volt and 15-20 ampere.

Fig. 16.1: Solar photovoltaic pumps at experimental fields of Central Arid Zone Research Institute, Jodhpur, (a) solar pump with 1 hp AC motor and 7×200 W_p PV array, (b) solar pump with 1 hp DC motor and 4×230 W_p PV array.

Solar irradiation vs pump performance

Since the availability of solar radiation varies in a day and also in different seasons of a year the performance of solar PV operated pumps also varies accordingly. In a field experiment, solar irradiation on horizontal surface has been measured continuously using radiation sensor of a portable weather station and simultaneously the pump discharges were recorded at intervals. Solar radiation of 400-500 W m^{-2} has been observed during early morning and late afternoon, whereas peak radiation reached up to 800-900 W m^{-2} during mid noon time. On a tilted surface, this available radiation will be slightly higher. Measured pressure-discharge relationship of both 1 HP solar AC and DC pumps with a suction head of 4-5 m revealed a maximum discharge of 120-140 l min^{-1} (lpm) at full radiation. Operating pressure has been observed as 1-1.2 kg cm^{-2}. It has been found that in case of solar AC pump, output voltage dropped to a level of 120-130 V during cloud shading, and thus the pump stopped to lift water. Moreover, during early morning hours and late evening hours AC output from PV array was not sufficient to start the pump. However, in case of DC solar pump, the effect of low irradiation on pump performance was minimum and has been found to start early in morning during winter months (~8:30 am) and continues till late evening (~5:30 pm) however, the discharge was low during these periods. Overall, it has been observed that DC solar pumps could be operated for longer period in a day than AC solar pump. Effect of solar radiation on pump discharge in the 1 HP AC solar pumping systems is shown in Fig. 16.2. Here, in a single day on 8th July 2014, pump discharge was measured four times with different solar radiation intensity but at similar head (7.05 ft). It was observed that at 14.2 psi which is equivalent to 1 kg cm^{-2} or 100 kPa, pump discharge was about 60-70 l min^{-1} from morning to noon when radiation was about 700-850 W m^{-2}, however it dropped to about 10-15 l min^{-1} during afternoon when the irradiation was about only 470 W m^{-2}. With low operating pressure, pump discharge at noon time was much higher (200-220 l min^{-1}) than morning and afternoon time (120-150 l min^{-1}). Therefore, it is observed that more is the available solar radiation, greater is the pump discharge and the pump performance was good. Here it is noted that during noon time the pump discharge at 10-15 psi (70-110 kPa) was slightly lower than morning time although the available radiation was higher. This is due to the increased PV panel temperature during noon time, which has a direct effect on reduced voltage output and generally performs optimally at 25°C. Because the irradiance varies with the time of the day, the power available for the pump also varies with time and thus needs proper tracking towards sun. From tracking experiments it was observed that instead of fixed south facing PV array, if the PV panels are tracked thrice a day; east facing during morning time (up to 11:00 am), south facing during noon time (11:00 am to 2:00 pm) and west facing (after 2:00 pm) during afternoon time, the amount of radiation received by the panels will be maximum and thus the pump's performance will be higher (Singh and Pande 2000).

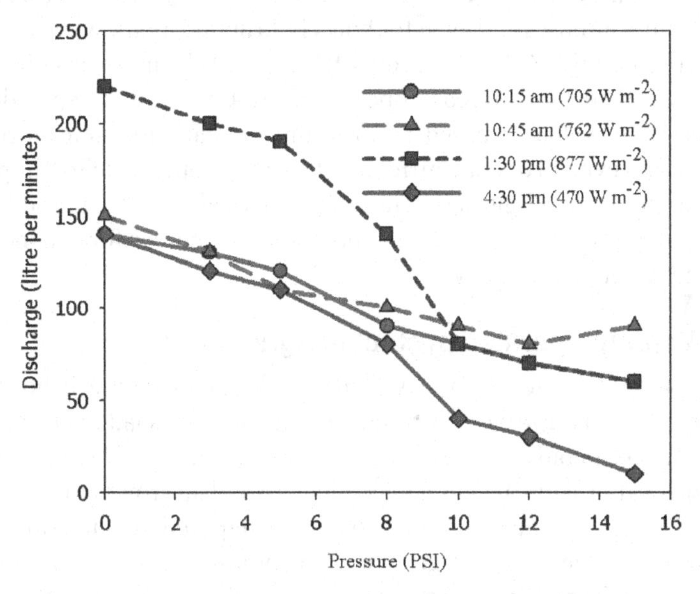

Fig. 16.2: Pressure discharge relationship of solar PV pump (1 hp AC pump with 1400 W_p PV panel)

Off-time utilization of solar pumps

The solar pumping system installed in a field are not always used for irrigation because crops require water at certain critical stages or irrigation water is applied at certain intervals or even there are some lag periods between two cropping seasons. During these off periods, PV panels of a solar pumping system continue to generate electricity but it is not used to operate the pump for irrigation and therefore the generated electricity gets wasted without any proper utilization. Hence, it is necessary to utilize the generated electricity somehow so that the system may become cost effective. In view of the need of creating facility for multipurpose use of solar pumps, a changeover switch in AC solar pump system has

Fig. 16.3: Change over switch in solar AC pump for off-time utilization of electricity produced by PV array

been developed. In this system one MCB switch is attached between DC-AC converter and AC pump. This will enable the farmers or user to utilize the electricity generated by PV array of solar pumps for different farm mechanization purposes and even for household applications, when the pump is not in operation (Fig. 16.3). The utility of developed change over switch system was successfully demonstrated to operate fan motors of 59 W capacities in the earth tube heat exchange based temperature regulation system in protected agriculture and even to operate different electrically operated farm implements such as ber grader and lighting systems in farm households. This addition has far reaching implications in wider adoption of solar pumps for enhancing both energy and water productivity.

Solar PV pump operated protected agriculture

The solar pumping system is mostly used with dripper networks in horticultural production systems, for which protected structure e.g. shade net house, poly house etc are commonly used. However, in arid and semi-arid region, heat load inside protected structure makes it uncongenial for plant growth for most periods in a year. Therefore, suitable cooling system is required to maintain the inside temperature suitable for plant growth. The cooling pad system is commonly used, which again required scarce water to run the system. Keeping in view the cooling requirement during hot summer months and scarcity of water for operating evaporative cooling system, design of earth tube heat exchange based protected agriculture system has been prepared, which could be operated with solar pumping system. For regulating the inside temperature of this proposed protected agriculture system, earth tube heat exchange pipes of 6" dia and 40 m long have been laid at 1.2 m depth below ground, which were further attached with a exhaust fan (58 W capacity) at outlet (Fig. 16.4). In this system, ambient air entered through inlet of the piping system passes through 40 m length embedded below ground and thus gets cooled during summer and warmed during winter months. The modulated air ultimately is blown inside the protected structure through exhaust fans, fixed at outlet of the embedded pipes. Initial observations have shown reduction of air temperature inside the protected structure by 4-5°C during late summer months. Air temperature and wind speed at outlet of the embedded earth tubes were recorded as 30°C and 3.5-4 m s^{-1}, respectively while the ambient temperature was 34-36°C. Whole protected agriculture system was successfully operated with changeover switching facility developed in the 1 hp AC solar pumping system. The developed system may be useful for small sized solar pumping system attached with rain water harvesting systems in surface ponds.

Fig. 16.4: Protected agriculture system with earth tube heat exchange systems with a fan motor operated by solar PV pumping system (1 hp AC pump).

Field performances of solar PV pump operated irrigation systems

Solar PV system is recommended to be used with efficient irrigation methods e.g. drippers and sprinklers etc. The solar pumping systems available in market with 3 HP and 5 HP capacities are expected to generate sufficient pressure to lift groundwater from a depth of about 75 ft to 200 ft and to operate drippers and sprinklers. However, considering the scarce groundwater situations, surface rain water storage systems with small sized solar pumping based irrigation is thought of as an optimal solution for future water use. Field experiments at Jodhpur, Rajasthan revealed satisfactory performance of dripper, micro-sprinkler and mini-sprinkler under 1 hp capacity solar pumping system (Fig. 16.5).

Performance of micro-irrigation system under the 1 hp capacity AC solar pumps with 3-4 m suction head revealed 2.0-2.2 kg cm^{-2} operating pressure, which successfully ran 9 mini-sprinklers with radius of throw of about 6-8 m. Similarly, the 1 hp capacity DC pump generated an operating pressure of 1.1-1.5 kg cm^{-2}, which ran 50 micro-sprinklers with throw radius of about 2 m and 160 drippers (capacity: 4 l h^{-1}) distributed in 8 laterals with 1 m spacing between drippers. Discharge of 45-50 litre per minute has been observed with 9 mini-sprinklers in the solar AC pumping system. Continuous measurements of radiation vs pump performance revealed that that DC solar pump may be operated for longer period in a day than AC solar pump. Dust load on PV array and its effect on pump performance has also been monitored and it has been observed that regular cleaning of PV panels is essential for optimum performance of solar pumps.

Micro-sprinklers Drippers

Mini-sprinkler system Mini-sprinkler

Fig. 16.5: Micro-irrigation systems under solar pumping system

Summary

Field performance of a 1 HP capacity solar PV pumping system is discussed here. For operating a 1 HP AC pump about 1400 W_p solar PV module array is required whereas for operating a DC pump of same capacity, 900 W_p solar PV module array is required. The pup performance at different solar irradiation from morning to afternoon is discussed and it was observed that the pump could be operated for about 6-7 hours a day at Jodhpur. The off-time utilization of the pumping system was also demonstrated using a change-over switching facility. Further, the pumping system was demonstrated to operate earth tube heat exchange based temperature regulation facility inside a protected agriculture production system. The pumping system was also found suitable for operating drippers, micro-sprinklers and mini sprinklers with an operating pressure of 1-1.5 kg cm^{-2}. From these observations, it is concluded that small sized solar PV pump e.g. 1 HP capacity can be used for irrigating crops through drippers inside protected cultivation system and even for applying irrigation water through sprinklers in an area of about 1/5th ha or about 1.5 bigha.

17

Phase Change Material Based Solar Dryer for Value Addition of Agricultural Produce

Dilip Jain and Soma Srivastava

ICAR- Central Arid Zone Research Institute, Jodhpur, Rajasthan, India

Introduction

Agricultural produce are being sun dried since ages as the first step to preserve the product for future consumption. It is still practised being most inexpensive over mechanical and other advanced drying systems. The most prevailing disadvantage of open sun drying is the quality losses due to insect infestation, enzymatic reactions, microorganism growth, and mycotoxin development. On the other side, sun drying is also a highly labor intensive and time consuming process prone to theft and damage by birds. It also suffers with the lack of process control and treatment uniformity (Bansal and Garg 1987, Bansal, 1987).

To overcome these disadvantages several types of solar dryers have been developed over the years (Ekechukwu and Norton 1999). These dryer may be identified into three group namely active, passive and hybrid. As regards their structural arrangement three generic subclasses were also identified: direct-modes (the solar-energy collection unit is an integral part of the entire drying system), indirect-modes (the solar collector and the drying chamber are separate units) and mixed-modes solar dryers. Natural convection solar dryers have become more suitable for rural sector and remote areas, as they work on single energy option and do not require any other external energy source (Pangavhane et al. 2002).

A continuous process is required while drying of crops until it reaches a desired moisture content that is not possible with solar drying after sunshine hours.

Thus, for continuous drying, a thermal storage could be provided with the solar air heater (Abul-Enein et al. 2000). A thermal storage unit integrated with the solar air heater may be charged during the peak sunshine hours and utilized (discharged) during off sunshine hours for supplying the hot air to the dryer (Jain and Jain 2004, Jain 2005a, 2005b). Heat storage using 'phase change materials' is a wise alternative that results in a continuous process throughout the day and night. More recently Tyagi et al. 2012, classified solar air heater on the basis of energy storage, numbers of covers, extended surface and their tracking axis.

Herbs are generally used in either its fresh form or dried form. Drying of herbs needs to be carried out under controlled conditions at low temperature so that flavor and color can be preserved. Thus, as an alternative to mechanical dryers, a solar dryer with controlled drying process can be utilized. In case of herbs and spices, the role of the dryer is not to dry more quickly, but to give a better quality product. The direct exposure to sun rays can greatly reduce the quality of herbs thus to prevent such losses an inclined flat-bed solar dryer with phase change thermal collector was designed and developed. The basic aim behind this research was to develop a crop dryer that can be helpful for drying those products also that extend for longer time of drying to reach desired moisture content. Development of solely solar dryer by using phase change material as thermal energy storage will provide continuous and uniform drying of spices and herbs for better quality products. The unique property of phase change material is storage of latent heat that can be used for thermal energy storage for continuous operation after sunshine hours during solar drying.

Solar drying with thermal energy storage

Conservation of available solar energy is needed to improve the efficiency of dryer and production of higher quality products. Solar energy is abundantly present in country like India during day time that can be stored for later use during night. Energy storage has a prime role in saving of fuels and also leads to a more cost effective system reducing wastage of energy and capital cost (Sharma et al. 2009). A thermal energy storage system is needed for judiciary utilization of available solar energy. Two known methods for thermal energy storage are sensible heat method and latent heat method. Latent heat storage method provides much higher storage density, with a smaller temperature difference between storing and releasing heat. (Farid et al. 2004). The basic work for thermal energy storage is done with an aim to store surplus heat during sunshine hours for later use during night (Jain 2007).

In some recent studies, the natural convection dryers have been developed for drying the fruits (Pangavhane et al. 2002, EL-Sebaii et al. 2002) and could be very well used during sunshine hours.

Phase change material (PCM)

Phase change materials are integrated in solar dryer with an aim to store energy and have an enormous potential to increase the effectiveness of energy conversion equipment use and for facilitating large-scale fuel substitutions. All those materials that have a large latent heat and high thermal conductivity can be a potential source for energy storage. Another most important property of PCM that plays an important role in choice for energy storage material is melting point. Since PCM is based on latent heat method it stores energy by change in its phase. Several types of PCM have been studied over the years such as paraffin wax (Faridand Mohamed 1987), fatty acids (Feldman et al. 1989), palmitic acid (Hasan 1994), hydrated salts (Biswas 1977) etc.

PCM uses chemical bonds to store and release heat for transfer of thermal energy that occurs when there is a change of state of PCM as they absorb solar heat. During phase change large amount of heat is stored that is released during off sunshine hours that helps in maintaining a higher temperature inside dryers compared to ambient temperature. Encapsulation of PCM results in large heat transfer area, reduction of the PCMs reactivity towards the external environment and controlling the changes in volume of the storage materials as phase change occurs (Farid et al. 2004). A solar air heater embedded a phase change material (PCM) heat storage system is suitable for crop drying applications (Enibe 2002).

Solar dryer details

An inclined flat plate collector was provided with glazing-I above an absorber-I and insulation material below. Air inlet is provided between glazing and absorber plate. A drying setup with three chambers is placed after collector in series (Fig.17.1). In the lower chamber of drying setup, a packed bed of phase change material (PCM-I) is placed in cylindrical tubes (all of same dimensions). Above the packed bed (PCM-I) drying trays are arranged in three rows and two columns to keep the sample to be dried. An inclined absorber plate-II attached with the storage material (PCM-II) of cylindrical tubes is placed over the drying plenum. A certain gap is provided between thermal storage (PCM-II) and absorber plate-II. Another reflector-II is placed in front side of the plenum as given in Fig.17.1. The working principle of flat plate collector is similar to as usual in conventional collector in addition of drying set up.

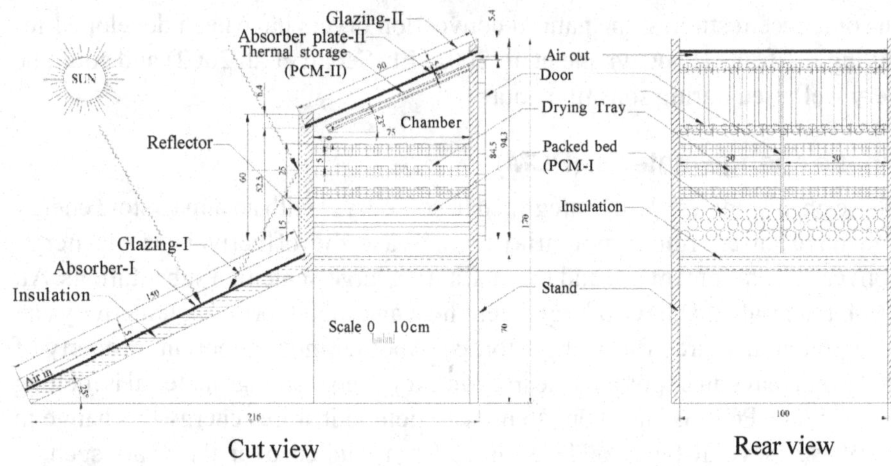

Fig. 17.1: Schematic view of flat plate absorber with thermal storage natural convective solar crop dryer

Working Principle of Dryer

The system is assumed to face towards the midday sun. As the radiation will fall directly on absorber plate-I and also reflected through reflector-I to absorber plate-I results in maximum entrapment of radiations, while the glazing-I will prevent thermal losses. The air heats up and flows through gap provided in between absorber plate-I and glazing-I and drying chamber will get heated by convection. The flowing air creates a draft due to thermal buoyancy and air starts flowing in the system. The higher temperature of air is utilized in heating and melting of PCM as thermal storage and remaining is passed over to the drying trays, which will start drying the crops during sun shine hours. The moist air from drying trays will move upward and when it will come in contact with absorber plate-II, which again gets heated and help in creating a draft due to thermal buoyancy and thus moisture laden air will be released out of the system. An inclined absorber plate-II attached with the storage material (PCM-II) in cylindrical tubes is placed over the drying plenum to build the thermal buoyancy for natural convection of moist air during off sunshine hours. During off sunshine hours the heat stored in PCM will be released, as the PCM will get solidified, that will release latent heat. This latent heat is utilized by crops during off sunshine hours to continue drying at least for 5 to 6 hours that will enhance the drying of crops. The basic aim to extend the drying hours is to reduce the time, utilize maximum available energy and prevent microbial and other physico-chemical losses due to left moisture.

Description of Dryer

A solar dryer with inclined flat plate absorber with thermal storage natural convection was designed (Fig 17.2). The solar dryer is divided into four major parts i.e. flat plate collector, packed bed for thermal storage, drying chamber and natural draft system.

Fig. 17.2: A flat plate absorber with thermal storage natural convective solar crop dryer

i. **Flat plate collector:** A 1.5 m flat plate with insulation at base and inclined at an angle of 25^0 C was attached in front of the drying system. The collector has a toughened glass that was placed on the flat plate at a distance of 0.05 m. This space in between allows the air to flow inside the dryer. The air gets heated as it travels from inlet to the drying system by the solar heat. The plate was painted with matt finished black color that further helps in heating of collector.

ii. **Packed bed:** A PCM thermal energy storage system was placed below the drying chamber that consisted of 48 numbers of cylindrical tubes. The tubes were of 0.75 m in length and had a diameter of 0.05 m. The tubes were filled with PCM and tightly packed to avoid any leakage. The tubes were placed in zigzag orientation such that around 48 kg paraffin wax material was stored that gets melted during day time and stores energy in the form of latent and sensible heat. This stored heat is utilized during night time for drying of crops.

Fig. 17.3: Inner view of dryer showing PCM packed bed thermal storage and drying trays

iii. **Drying system:** A drying system with six rectangular drying trays of dimension 0.50 m × 0.75 m was placed above packed bed in three rows and two columns. The drying tray with stainless steel frame and mesh was prepared for drying of samples. The trays have a total capacity of 12-16 kg. Totally six trays were placed in two parallel rows.

iv. **Natural draft system:** A toughened glass was placed on the top absorber plate. The vent present at top allows moisture laden air to move out of the drying chamber. This air passage system helps in movement of air. A packed bed with PCM is placed near the vent to heat the air and enhance its movement.

v. **Reflector mirror:** A reflector mirror is placed facing south adjacent to drying chamber. The purpose is to enhance the heating of air as it reflects most of the falling radiations to flat plate collector below. This additional falling radiation further helps in improving the efficiency of dryer.

Experimental Parameters for Optimization of Drying Process

The solar dryer having the two level of PCM thermal energy storage i.e. at hot air inlet in drying chamber and another above the drying chamber and outlet air passage (Fig. 17.3). A PCM-I thermal energy storage system was placed below the drying chamber that consisted of 48 numbers of cylindrical tubes and PCM-II of 32 numbers placed above the drying chamber as to create natural convection. The tubes were filled with paraffin wax and tightly packed to avoid any leakage. The tubes were placed in zigzag orientation such that around 48 kg paraffin wax material was stored that gets melted during day time and stores energy in the form of latent and sensible heat. This stored heat was utilized during night time for drying of crops. A natural draft system was inbuilt above the drying chamber with an absorber plate at an inclination of 23°. A toughened glass was placed on the top absorber plate to prevent top losses. Second packed bed with PCM was placed 0.05 m below the absorber plate (Fig. 17.2), which provided the vent space for movement of heated air and enhanced convection. The experimental optimization was conducted with four combinations.

Test-1 Without PCM-I and PCM-II

Test-2 On row (16 tubes in each) of PCM-I and PCM-II

Test-3 Two row (32 tubes in each) of PCM-I and PCM-II

Test-4 Three row (48 tubes) of PCM-I and two row (32 tubes) of PCM-II

A flat plate PCM packed bed solar dryer was installed in solar yard of ICAR-CAZRI, Jodhpur facing south with an aim to utilize maximum falling radiations for drying. Data logger was attached with solar dryer to record the temperature at various stages of dryer. A mini weather station was installed simultaneously near the dryer for recording solar radiation and ambient temperature. Solar radiation, ambient temperature and RH were recorded in the weather station and similarly various temperatures and RH within the dryer were recorded in data logger at the interval of 15 min. The typical representation of the data recorded on solar dryer and weather station are presented in Fig. 17.4. The effect of thermal energy storages for each day is encircled.

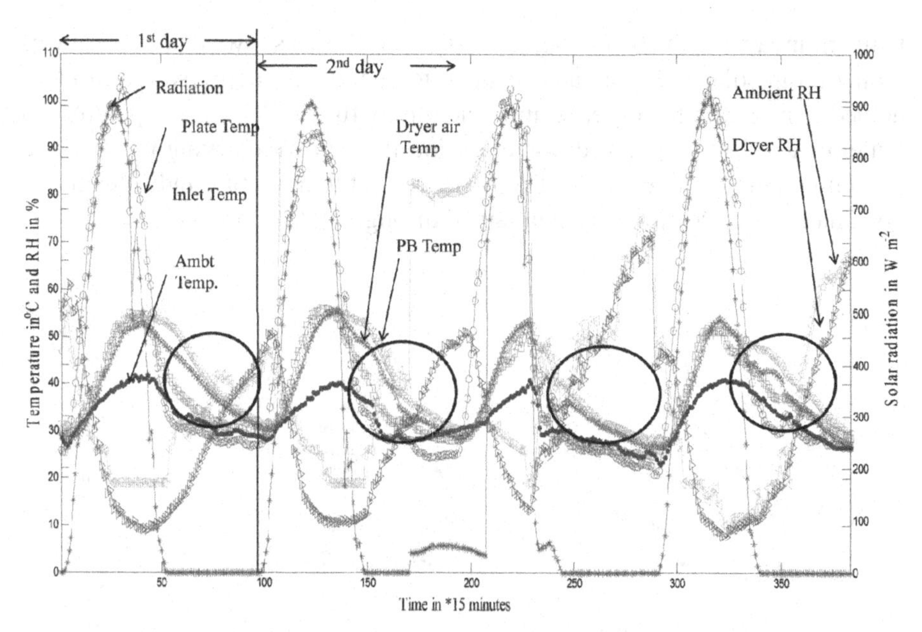

Fig. 17.4: Typical representation of solar radiation, ambient temperature and solar dryer temperature at various stage recorded for four days at 15 minutes interval in September

Overall Thermal Performance of PCM Fased Dryer

The overall thermal performance of solar dryer is presented in Fig. 17.5. The hourly mean values were obtained of each hour for graphical representation. Fig. 17.5 shows a diurnal variation of ambient temperature, solar radiation and dryer temperatures with respect to hourly time. The graph plotted was hourly means from 6:30 AM in the morning to 5.30 AM the next morning. The solar radiation initially at 6:30 AM was observed to be near 10 Wm^{-2} while with time it reached 900 Wm^{-2} at 1.30 PM. It declined afterwards and reached its minimum at 7:30 PM. This shows a trend of available solar radiation during a day. Simultaneously temperature is plotted against the time that reveals that it varied from 30 to 45°C during the day time. A maximum temperature was observed during afternoon time i.e. near 2.30 PM.

A very high flat plate temperature was observed that increased during day up to 95°C and then declined after sunset to ambient temperature. Plate temperature is an important factor in working of dryer as the inlet air will get heated from here and thus will help in drying process within the chamber. The temperature below and above the PCM bed was 20 °C higher than the ambient temperature. Although the temperature declined after 1.30 PM, it was still higher than ambient throughout night that shows an extended drying process during off sunshine hours. During initial hours of day the temperature below packed bed was higher compared to temperature above it that helps in charging the PCM. While temperature above the bed was higher after 2 PM, this shows that as there was reduction in solar radiation and ambient temperature, energy stored in PCM helped to maintain the dryer temperature up to 40 – 45 °C. Air temperature in outlet channel was observed higher during the complete drying process and maximum being above 50 °C. The higher outlet temperature indicates that a good natural convection was maintained throughout the process.

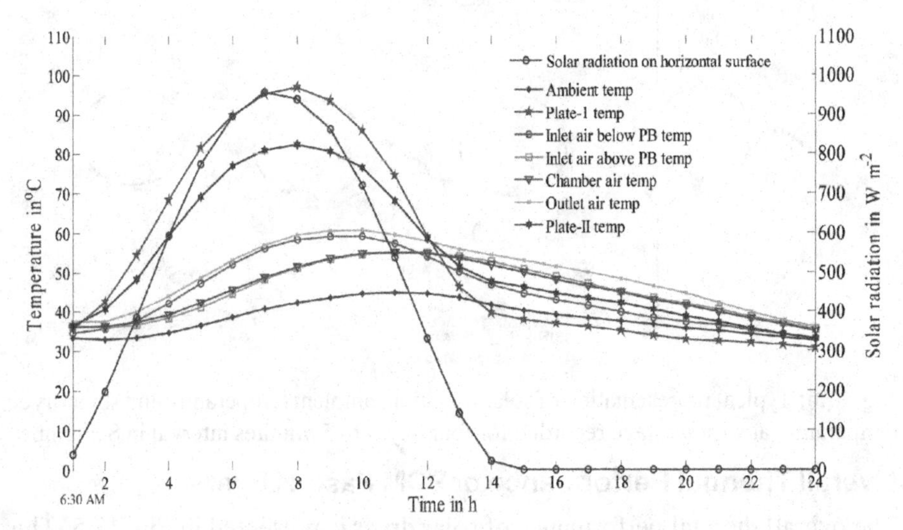

Fig. 17.5: Diurnal variation of temperature at various stage of flat plate PCM packed bed solar dryer during June in Jodhpur

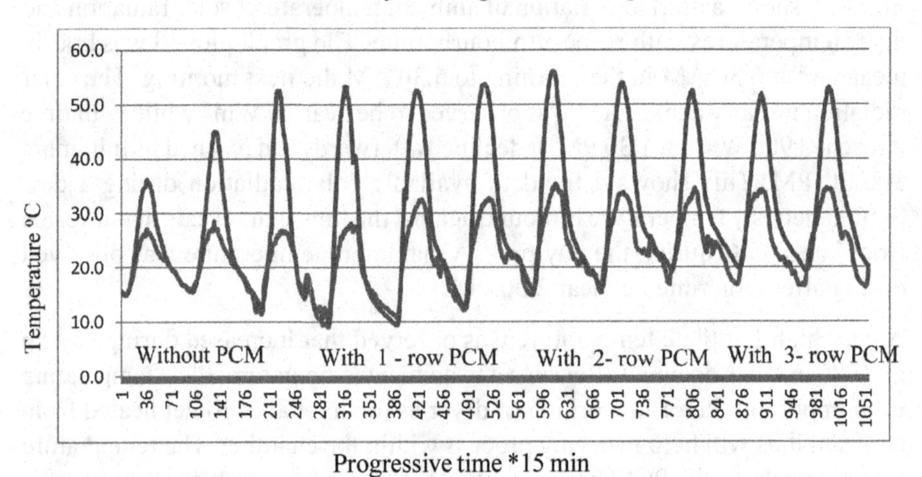

Fig. 17.6: Typical presentation of effect of various load of PCM on temperature of drying chamber

Effect of various load of PCM on temperature of drying chamber

Experiments were conducted as described in previous section for each load of PCM for consecutive 2 to 3 days in each test conditions. The temperature trend of drying chamber with ambient at every 15 min interval is shown in Fig. 17.6. It shows a representative performance of PCM application in solar dryer. By increasing the load of PCM show differential in increase in chamber temperature from the ambient after sun set hours.

These experiments were conducted from June to November, 2015. The hourly mean temperatures of drying chamber for 24 hours (6:30 AM to 5:30 AM next morning) for each test condition are presented in Fig. 17.7 and 17.8 for the months of June and November, respectively. These two months are representative of clear sunny days of summer and winter. The average ambient temperature of four days on hourly mean basis is also shown in these figures to view the relative effect of thermal energy storage.

The experimental condition in summer resulted in higher drying chamber temperature as 60 °C without PCM (Test-1) and lower after sunset hours (Fig.17.7). The variation in temperature during day and night of drying chamber can be observed more as from 60 to 30 °C. While using the PCM, the single row (Test-2) of tubes as thermal energy storage dropped the peak temperature in drying chamber, but the gain of thermal energy could not be observed after sun set as the temperature of drying chamber did not get much higher than the ambient. However, with using the more rows of PCM as Test-3 and Test-4 the higher temperature in drying chamber from ambient could be observed after

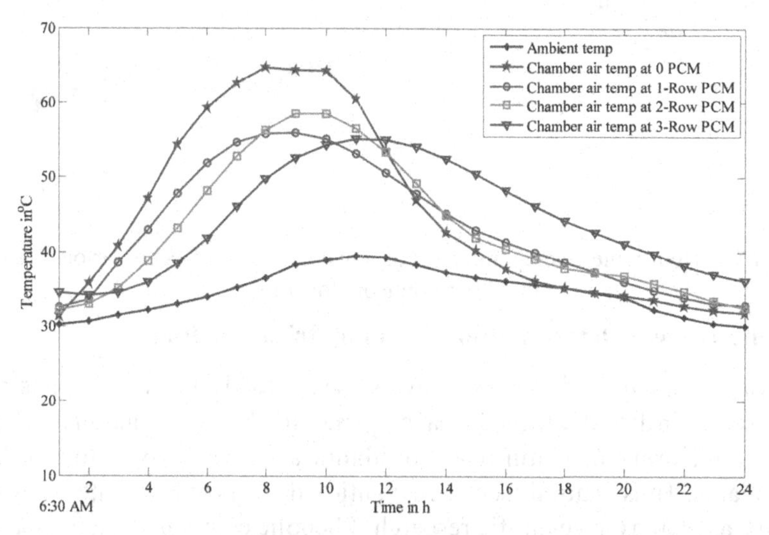

Fig. 17.7: Hourly mean temperature of drying chamber as effect ofvarious load of PCM for month of June

sun-set signified the effect of thermal energy storage. The variation in temperature during day and night of drying chamber can be observed less as from 55 to 38°C.

For the winter season condition, though the ambient temperatures (Fig. 17.8) are lower, however, the trends of working of the dryer was more or less similar as described earlier. The maximum variation of chamber temperature from ambient without PCM (Test-I) could be observed from 58 to 32°C. While increasing the load of PCM from single row to two row (Test-2 and Test-3), the chamber temperature being drop as 6-8°C and the uniform thermal can be observed after sun-set hours. However, putting the three row of PCM (Test-4) drops the chamber temperature during day time from 55 to 42°C, but did not yield much thermal storage after sun-set hours. Under this situation the combination of two row of PCM (Test-4) would have been kept as optimum thermal storage for winter season.

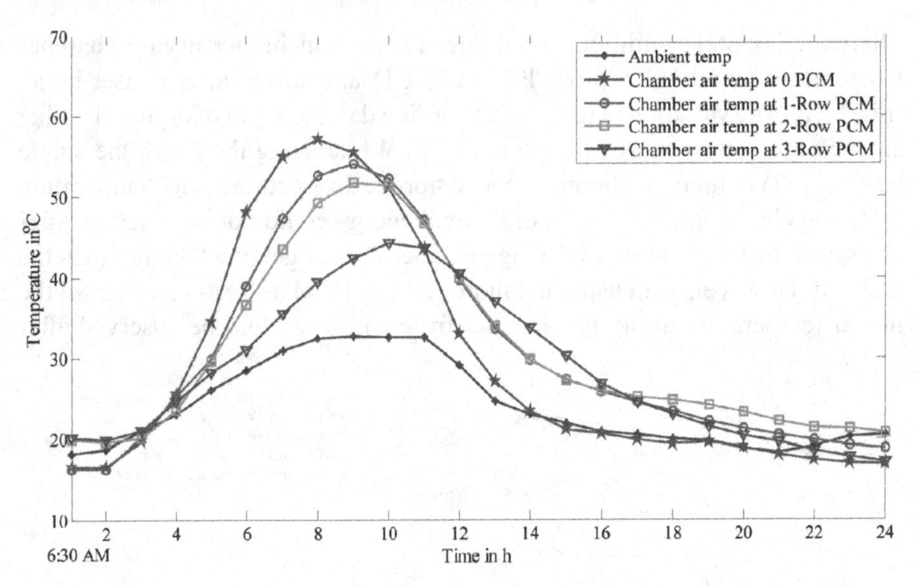

Fig. 17.8: Hourly mean temperature of drying chamber as effect ofvarious load of PCM for month of November

Phenolic content determination of solar dried products

Spices and aromatic herbs are extensively used as food additives are considered to be essential in diets due to medicinal properties for delaying aging and biological tissue deterioration. Natural antioxidants are alternatives to synthetic antioxidants. Thus, naturally occurring antioxidants is of great interest both in industry as well as in scientific research. Phenolic compounds have potential against oxidative damages, therefore play a protective role because of their antioxidant activities, they widely used in processed foods as a natural antioxidant. Three aromatic herbs e.g. coriander, mint, and fenugreek and spice crop e.g. green chilli were dried in the PCM based solar dryer, whereas open sun drying of these products were carried out simultaneously. Total phenols values were expressed in terms of Gallic acid equivalent (mg g^{-1} of dry extract).

Table. 1: Retention of total phenolic compound

S.No.	Sample	Fresh, gm GAE/100 gm [DM]	Sun Drying, (%)	Solar drying, (%)
1.	Coriander	9.010	6.050(67.14)	6.506(72.20)
3.	Fenugreek	9.171	7.949(86.67)	(8.427)(91.88)
2.	Mint	10.609	9.231(87.01)	9.583(90.32)
4.	Green chillies	8.990	6.761(75.20)	7.403(82.34)

Research has confirmed that these herbs possess total phenolic contents which are of enormous benefits. The antioxidant support we get from investigated herbs is largely due to their phenolic compounds that have been shown to help protect us against unwanted oxygen damage to our cells, blood vessels, and organ systems. Antioxidant compounds in food play an important role as a health protecting factor. Results showed that drying is more effective in retention of phenolic compounds than open sun drying (67.14) as solar drying was having 72.20 per cent retention in fenugreek. The trend was found similar for the other herbs also. Highest amount of total phenols was observed in fenugreek followed by mint > green chillies> coriander.

Scientific evidence suggests that antioxidants reduce the risk for chronic diseases including cancer and heart disease. It is increasingly being realized that many of today's diseases are due to the "oxidative stress" that results from an imbalance between formation and neutralization of pro-oxidants. Oxidative stress is initiated by free radicals; hence, all human beings protect themselves against free radical damage somehow by antioxidant supplements, which are vital to combat oxidative damage. It is also called free radical scavenging activity because of its ability to scavenge free radicals. Free radical generation is directly related with oxidation in foods and biological systems. Therefore, the search for potential free radical scavenging agent is imperative and indispensable.

Economic analysis of PCM based dryer

The solar dried products are considered of the quality and hygiene at par with branded products (Sreekumar 2010). Economic analysis of the solar devices is important to understand the application from the commercial point of view (Fudholi et al. 2011). Table 17.2 presents the values of various input parameters used in the financial evaluation of solar dryer.

Solar dryers are found to be techno-economically a suitable device for food drying (Jain et al. 2004). In this analysis return of capital, profit, simple payback period and net present value were calculated. The acceptability of the solar technology at commercial level increases with the short payback period (Seveda et al. 2004, Sukhatne 1998). Economic indicators were considered as annual profit, return on capital, simple payback period (SPB) and net present value (NPV) for feasibility analysis.

Table 17.2: Parameters for Economic Analysis

Descriptions	Value
Cost of system (C),₹	1, 00,000
Life of solar dryer (n), y	10
Capacity of dryer (S), kg d^{-1}	12
Discount rate (i), %	10
Cost of Insurance (I_r), ₹	0.1*C
Operational cost (C_o), ₹	0.1*C
Annual maintenance cost (C_m), ₹	0.01*C
Cost of raw product (C_r), ₹ kg^{-1}	15, 20, 25
Yield of dried powder (Y), kg	0.12*S
Packaging cost (P_k), ₹	$2*Y*d_n$
Cost of dried product (P), ₹ kg^{-1}	300, 400, 500
Wage of 2 man-h. d^{-1} (W_a), ₹	25
Labour cost of an year (L_w), ₹	W_a*d_n
Days of operation in year (d_n), d	240
Drying time of one batch (t_d), h	24
Salvage value of the dryer (S_v), ₹	0.15*C

The cost of the dried product (P) can be written as follows

$$P = C_r + C_o + C_m + L_w + P_k + I_r \quad (1)$$

where, C_r= Cost of raw material

C_o= Operational cost

C_m = Maintenance cost

L_w = Annual labour cost

P_k = Packaging cost

I_r = Insurance cost of dryer

The performance of an investment can be studied from the production. Profit is defined as the difference between total sales with all types of spending. Profits (P_r) can be written as

$$P_r = P_s - P \quad (2)$$

where, P_s is total sale price of product.

Return of capital (R_c) is also called the profit from the investment and is influenced by time, can be expressed as the ratio of profit to the cost of system (C).

$$R_c = \frac{P_r}{C} \tag{3}$$

The simple payback (P_b) is the investment cost per average annual net income. Returns are easy to recover the invested capital. Simple payback period is calculated as

$$P_b = \frac{C}{P_r} \tag{4}$$

Parameters of the rate of return of capital and the payback period are not easy to take into account the impact on value for money. While the net present value method (P_N) is calculating the present value of excess cash flow during the project period. NPV is the total present value for each year of the net cash flow less capital costs. It is calculated as follows

$$P_N = \Sigma P_n (1 + i)^n - C \tag{5}$$

where, $P_n = S(1 + i)^{-n}$ is the discounted present value (S) to the invested in the n years in the future. The considerable unit price of the raw material at 15, 20 & 25₹ kg^{-1} and the product unit sales price as 300, 400 & 500₹ kg^{-1} were taken for economic analysis. The matrix of different trend of economic indicators is shown in Table 17.3. It indicates that at the reasonable price of raw material 20₹ kg^{-1} and product cost price 400₹ kg^{-1} estimated the annual profit of ₹ 65,949 and payback period found to be 1 year 6 months.

Table 17.3: Economic indicator for solar dryer at various cash inflow and out flow

Cost of raw material,₹ kg^{-1}	Product Sales price,₹ kg^{-1}	Profit,₹	Return on capital	SPB, y	NPV,₹
15	300	45,789	0.4579	2.18	2,75,570
	400	80,349	0.8035	1.24	4,87,930
	500	1,14,910	1.1491	0.87	7,00,280
20	300	31,389	0.3139	3.18	1,87,090
	400	65,949	0.6595	1.51	3,99,440
	500	1,00,510	1.0051	0.99	6,11,800
25	300	16,989	0.1699	5.88	98,606
	400	51,549	0.5155	1.93	3,10,960
	500	86,109	0.8611	1.16	5,23,320

Table 17.4: Net cash flow of solar dryer for assuming the potential market price 400₹ kg^{-1} of product and 20₹ kg^{-1} of raw material, NPV = ₹ 3,99,447.5

Year	C,₹.	Net Cash Flow,₹	i, (10%)	Present Value	
				C,₹	P_n
0	1,00,000	65,949	1	1,00,000	
1		65,949	0.9091		59953
2		65,949	0.8264		54503
3		65,949	0.7513		49548
4		65,949	0.6830		45044
5		65,949	0.6209		40949
6		65,949	0.5645		37226
7		65,949	0.5132		33842
8		65,949	0.4665		30766
9		65,949	0.4241		27969
10		65,949	0.3855		25426
	15,000			5782.5	405230

Table 17.3 shows the net cash flow of the solar drying system assuming the potential price of raw material 20₹ kg^{-1} and the market price of the product as 400₹ kg^{-1}. The NPV value obtained as ₹ 3,99,447 with an annual profit of ₹ 65,949. Thus, it can be inferred that the developed solar system is economically and commercially viable.

Conclusion

The PCM based solar dryer is an alternative to overcome the disadvantages of traditional open sun drying and utilization of maximum available solar radiations. A dryer with packed bed PCM was capable of storage of thermal energy in the form of latent heat and sensible heat during day time and release the same heat after sun shine hours. This stored energy helped in maintaining the drying temperature ranged between 40 and 45°C, thus extended the period of drying. The requirement of PCM during the winter season was less than the summer conditions for optimum thermal storage effect. Considering reasonable cost of raw material 20₹ kg^{-1} and price of product 400₹ kg^{-1}, the solar dryer was financially viable with payback period of 1.5 year.

References

Aboul-Enein, S., El-Sebaii, A. A., Ramadan, M. R. I., & El-Gohary, H. G. 2000. Parametric study of a solar air heater with and without thermal storage for solar drying applications. Renewable Energy, 21, 505-522.

Bansal NK, Garg HP. 1987.In Advances in Drying, Ed.; Hemisphere Publishing: New York, Solar Crop Drying, Vol. 4.

Biswas, D.R.1977. Thermal energy storage using sodium sulphate-deca-hydrate and water. Solar Energy19:99–100.

Ekechukwu, O.V. and Norton, B. 1999. Review of solar energy drying II: an overview of drying technology. Energy Conservation and Management 40 (6):615–55.

El-Sebaii A. A., Aboul-Enein, S., Ramadan, M. R. I., & El-Gohary, H. G. 2002. Experimental investigation of an indirect type natural convection solar dryer. Energy Conversion and Management, 43, 2251-2266.

Enibe, S.O. 2002. Performance of a natural circulation solar air heating system with phase change material energy storage. Renewable Energy27(1): 69-86.

Farid, M.M. and Mohamed, A.K., 1987. Effect of naturalconvection on the process of melting and solidification of paraffin wax. Chemical Engineering Communications 57:297–316.

Farid, M.M., Khudhair, A.M., Razack, S.A. and Hallaj, S.A. 2004. A review on phase change energy storage: materials and applications. Energy Conservation and Management 45: 1597–1615.

Feldman, D., Shapiro, M.M., Banu, D. and Fucks, C.J.1989. Fatty acids and their mixtures as phase-change materials for thermal energy storage. Solar Energy Materials, 18:201–16.

Fudholi A, Ruslan MH, Yahya M, Zaharim A, Sopian K. 2011. Techno economic analysis of solar drying system for seaweed in Malaysia. Recent researches in Energy, Environment and Landscape Architecture, 89-95.

Hasan, A. 1994. Phase change material energy storage system employing palmitic acid. Solar Energy, 52:143–54.

Jain, D. 2005a.Modeling the performance of greenhouse with packed bed thermal storage on crop drying application. Journal of food Engineering,71(2): 170-178.

Jain, D. 2005b.Modeling the system performance of multi-tray crop drying using an inclined multi-pass solar air heater with in-built thermal storage. Journal of food Engineering71(1): 44-54.

Jain, D. 2007. Modeling the performance of the reversed absorber with packed bed thermal storage natural convection solar dryer. Journal of Food Engineering78:637-647.

Jain, D. and Jain, R. K. 2004.Performance evaluation of an inclined multi-pass solar air heater with in-built thermal storage on deep-bed drying application. Journal of Food Engineering 65: 497-509.

Jain, D. and Tiwari, G. N. 2003.Thermal aspect of open sun drying of various crops. Energy, 28: 37–54.

Jain, N. K., Kothari, S. and Mathur, A.N. 2004.Techno-economic evaluation of a forced convection solar dryer. Journal of Agriculture Engineering 41: 6-12.

Mani, A.1981. Handbook of solar radiation data for India 1980. Madras: Allied Publishers Private Limited.

Pangavhane, D. R., Sawhney, R.L. and Sarsavadia, P.N. 2002. Design, development and performance testing of a new natural convection solar dryer. Energy 27: 579-590.

Seveda MS, Rathore, NS, Singh P. 2004. Techno economics of solar tunnel dryer- A case study. Journal of Agriculture Engineering, 41(3): 13-17.

Sharma, A., Chen, C.R. and Lan, N.V. 2009. Solar energy drying systems: a review. Renewable Sustainable Energy Review, 13(6–7):1185–210.

Sreekumar, A. 2010. Techno- economics analysis of a roof integrated solar air heating system for drying fruit and vegetables. Energy Conservation and Management, 51: 2230-2238.

Sreekumar, A., Manikantan, P.E. and Vijayakumar, K.P. 2008. Performance of indirect solarcabinet dryer. Energy Convers Manage, 49:1388–95.

Sukhatme, S.P. 1998. Solar energy. New Delhi: Tata McGraw-Hill Limited.

Tyagi VV, Panwar NL, Rahim NA, Kothari R. 2012. Review on solar air heating system with and without thermal energy storage system. Renewable and Sustainable Energy Reviews, 16(4): 2289-2303.

18

Application of PCM Materials in Temperature Regulation Inside Protected Agriculture System

Priyabrata Santra, P.C. Pande, A.K. Singh, S. Poonia

ICAR-Central Arid Zone Research Institute, Jodhpur, Rajasthan, India

Introduction

Cultivation of crops in greenhouse or protected structure has been increased tremendously in recent times and also has been spread from temperate regions to the warmer regions of tropics and subtropics. Plentiful technologies have been developed for heating the inside environment of greenhouse for optimum plant growth (Sethi et al. 2013). However, in hot arid areas, cultivation in greenhouse is very difficult as the temperature reaches to a very high level to sustain plant growth (Pek and Hayles 2004, Saran et al. 2010). Under such situation requirement for cooling the greenhouse microclimate environment is the necessity, but is very energy and water intensive. Hence, in such regions, reduction of air temperature inside the greenhouse or the regulation of temperature closer to the ambient temperature during summer is necessary for successful crop production. A typical diurnal variation of temperature and global solar radiation during late winter months inside a shade net structure has been plotted in Fig. 18.1. It has been found that global solar radiation has been decreased to almost 1/3 of the radiation received outside by using 75% shade net cover. The temperature inside structure during morning time has also been found higher than outside temperature and in summer such increase is expected to be much higher.

Therefore, cooling is considered as the basic necessity for greenhouse crop production in tropical and subtropical regions to overcome the problems of high temperatures during summer months. Development of suitable cooling system that provides congenial microclimate for crop growth is a difficult task as the design is closely related to the local environmental conditions. Different cooling techniques for greenhouse crop cultivation may be found in Kumar et al. (2009). Fan pad cooling system has been widely used for greenhouse cultivation however, requires large amount of water during evaporation based cooling. Since water resources are becoming scarce in future such fan-pad cooling system may not be viable in future specifically in hot arid region.

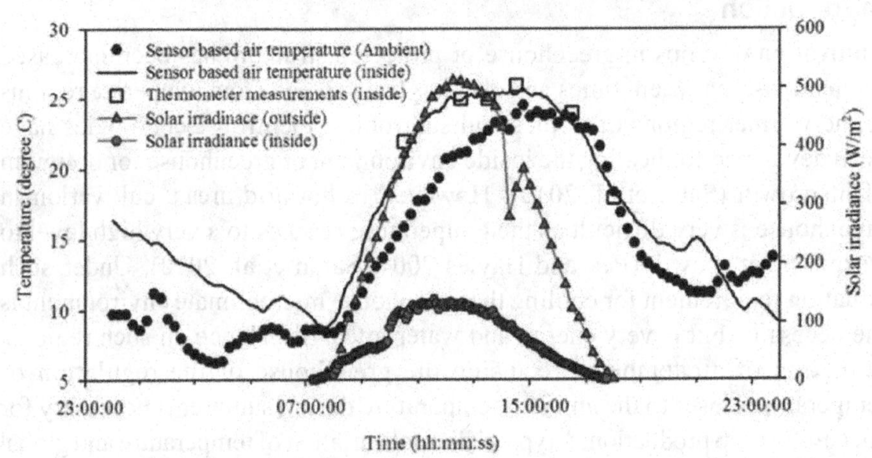

Fig. 18.1: Protected shade net structure for crop production and the inside environment recorded on a winter day at Jodhpur

Depending on the applications, the PCMs should first be selected based on their melting temperature lying in the practical range of operation, which is about 25-35°C for protected agriculture system. Materials that melt below 15 °C are used for storing coolness in air conditioning applications, while materials that melt above 90 °C are used for absorption refrigeration. All other materials that melt between these two temperatures can be applied in solar heating and for heat load leveling applications. Commercial paraffin waxes are cheap with moderate thermal storage densities (~200 kJ kg^{-1} or 150 MJ m^{-3}) and a wide range of melting temperatures. They undergo negligible subcooling and are chemically inert and stable with no phase segregation. However, they have low thermal conductivity (~0.2 W m^{-1} $°C^{-1}$), which limits their applications. Pure paraffin waxes are very expensive, and therefore, only technical grade paraffins can be used. Keeping in mind the cooling requirement of protected cultivation system and potential of paraffin wax as suitable PCM material for heat storage, it was tried to develop a material mixture of different grades of paraffin waxes and a heat exchanger for regulating the inside temperature of protected agriculture system at hot arid region of India.

PCM preparation

Mixtures of paraffin liquid and paraffin wax was prepared as PCM material at laboratory with different ration. For proper mixing of two materials, they were first heated and then mixed thoroughly using a stirring rod. Temperature changes of the prepared mixture material during heating and cooling process was determined using a water bath. During heating experiment, water in the bath and mixture material was heated from its normal room temperature and change of temperature of both water and the material was recorded using glass thermometer. In the cooling experiments, mixture materials were kept inside water bath, which was preheated to 75-80°C. Apart ftom these, hot water bath experiment at a predefined temperature (75-80°C) was also carried out using a thermostat system. Heat gain and heat loss during heating and cooling experiment was calculated by the following equation:

$$q = m.s\ \Delta t \qquad (1)$$

Where q = amount of heat loss or gain, m = mass of water, s = specific heat capacity of water and Δt = change in temperature.

After identifying a suitable mixture as PCM material with melting temperature at around 35°C, it was first tested inside a small prototype greenhouse structure and then in a PV clad enclosure. A heat exchanger of GI tank along with aluminium fins surrounding it was prepared and PCM material was stored inside it. The PCM was filled inside the tank up to 80-85% level and then kept inside prototype structure. Temperature of both inside and outside of the prototype structure was monitored at half hourly interval. Prototype structure was prepared using transparent PVC sheet at top and surround wall whereas the floor was prepared with plywood sheet. The performance of PCM was also tested inside a PV clad enclosure with chili crop grown inside. Sensible heat storage and latent heat storage of the heat exchanger system was quantified using the Eq (1). The specific heat capacity of the heat exchanger material was considered as 0.22 kCal kg^{-1} $°C^{-1}$ for aluminium, 0.122 kCal kg^{-1} $°C^{-1}$ for GI and 0.24 kCal kg^{-1} $°C^{-1}$ for air.

Characteristics of PCM

Typical temperature profiles of PCM material mixture are presented in Fig. 18.2. It has been observed that during phase change, temperature of PCM material remained almost similar for a period, which was about 32-35 °C. It was also noted that when phase changed occurred, physical state of PCM material changed from semi-solid to semi-liquid state.

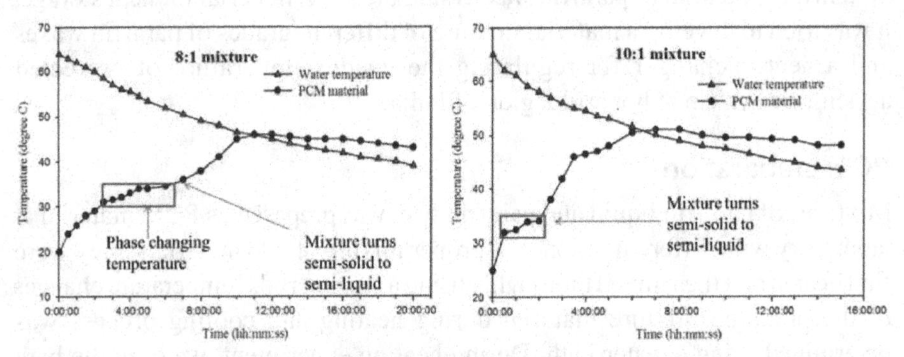

Fig. 18.2: Temperature profile of PCM material in water bath experiment with 8:1 and 10:1 mixture of paraffin liquid and wax

Similarly the temperature profile of PCM material during hot water bath experiment was presented in Fig. 18.3. Characteristically, it has been observed that temperature of PCM material mixture remained constant at about 34-35°C after 2 minutes from the start of experiment.

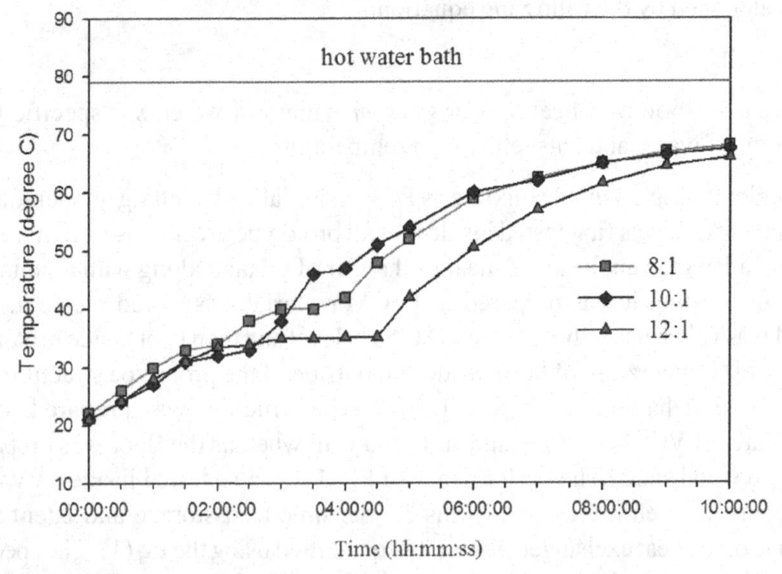

Fig. 18.3: Temperature profile of PCM material in hot water bath experiment with 8:1, 10:1 and 12:1 mixture of paraffin liquid and wax

Phase changing temperature

Identified phase change temperature of different PCM material mixture is given in Table 18.1. Mixture of paraffin liquid and wax at a ratio of 10:1 was found best suitable for regulating temperature of protected agriculture structure at about 35°C.

Table 18.1: Phase changing temperature of PCM material prepared with mixing of paraffin liquid and wax at different ratio as obtained from water bath experiment

Mixture of paraffin liquid and wax	Density (g/cm^{-3})	Phase changing temperature
6:1	0.841	37-39
8:1	0.808	40-42
10:1	0.765	31-35
12:1	0.861	34-35

Heat exchanger for storage of PCM

Two heat exchanger structures consisting of GI sheet tank of 6.5 litre capacity and aluminium fins attached at both sides of the tank for better heat transfer with air was designed and fabricated and these were tested inside small experimental structures with fibre glass sheet casing and covers (Fig. 18.4). Temperature profile throughout the day was recorded both inside and outside with PCM load and without load condition. It was observed that temperature inside structure without any PCM load reached to 60-62°C, when ambient was about 33-34 °C. With PCM load of 5.2 litre, inside temperature was found 53-54 °C, when ambient was 33-34 °C. Thus, a reduction in inside temperature by 7-8 °C was observed by using PCM material.

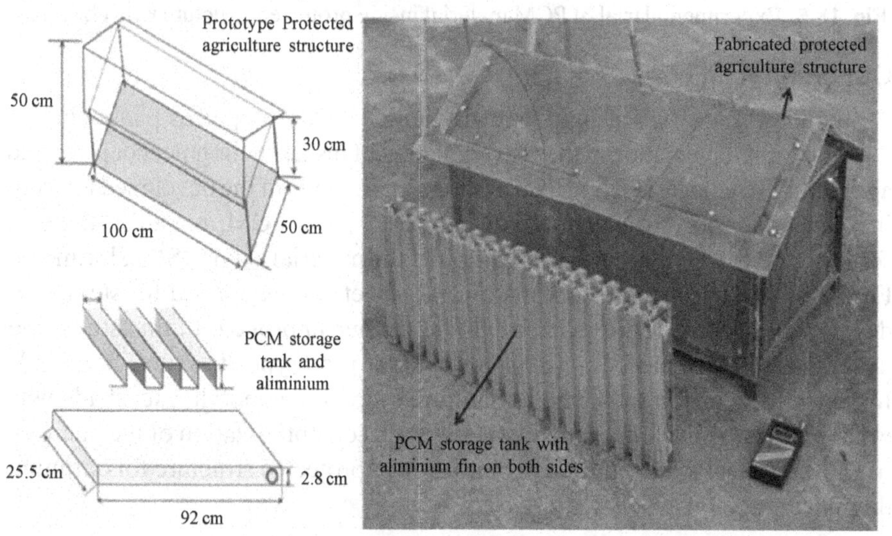

Fig. 18.4: Heat exchanger structure for storing PCM material and prototype protected agriculture structure for experiment

Testing of PCM material inside protected structure

Subsequently, the system was kept inside the PV clad enclosure and the performance of the enclosure was studied for growing chillies (Fig. 18.5). The amount of heat stored by a single PCM storage system with 2.5 kg PCM materials was found about 1100-1200 kJ. An estimate indicates that it is sufficient to reduce the inside temperature by 1°C for four hours if number of air change rate is 30 h^{-1}. However, long term trials are required to ascertain number of cycles of heating and cooling of the material. Results have so far indicated regulation of temperature within desirable limits leading to reasonable growth of chillies plants with good flowering and bearing fruits (Fig. 18.5).

Fig. 18.5: Experimental trial of PCM material inside protected structure with chili crop

Conclusion

A mixture of paraffin wax and paraffin liquid was developed as phase change material for its application in thermal regulation of inside environment of protected agriculture structure. Heating and cooling experiments of the developed mixture in laboratory showed characteristic properties of a PCM, however, there is further need for detailed characterization of the material using DSC calorimeter. Furthermore, a heat exchanger made of GI sheet was developed for storage of developed PCM material, which may be further improved. In this study, the developed PCM material was tested inside a prototype structure, but needs further testing inside protected agriculture systems e.g. polyhouse, shade nets et car required. Such experiments should focus on optimization of the quantity of PCM material to be required per unit area of protected structure for optimum thermal regulation inside it.

References

Sethi, V.P., Sumathy, K., Lee, C., Pal, D.S. 2013. Thermal modeling of solar greenhouse microclimate control: A review on heating technologies. Solar Energy, 96, 56-82.

Pek, Z., Hayles, L., 2004. The effect of daily temperature on truss flowering rate of ornamental crops. Journal of Science of Food and Agriculture, 84 (13), 1671–1674.

Sharan, G., Madhavan, T., 2010. Cropping in semi-arid northwestern India in greenhouse with ground coupling shading and natural ventilation for environmental control. International Journal for Service Learning in Agriculture, 5(1), 148-169.

Kumar, K.S., Tiwari, K.N., Jha, M.K., 2009. Design and technology for greenhouse cooling in tropical and subtropical regions: A review. Energy and Buildings, 41, 1269–1275.

Farid, M.M., Khudhair, A.M., Razack, S.A.K., Al-Hallaj, S., 2004. A review on phase change energy storage: materials and applications. Energy Conversion and Management, 45, 1597–1615.

Berroug, F., Lakhal, E,K., Omari, M. El., Faraji, M., Qarnia, H. El., 2011. Thermal performance of a greenhouse with a phase change material north wall. Energy and Buildings, 43, 3027-3035.

Harrou, H., Guilioni, L., Dufour, L., Dupraz, C., Wery, J., 2013. Microclimate under agrivoltaic systems: Is crop growth rate affected in the partial shade of solar panels. Agriculture and Forest Meteorology, 177, 117-132.

Khudhair, A.M., Farid, M.M., 2004. A review on energy conservation in building applications with thermal storage by latent heat using phase change materials. Energy Conversion and Management, 45, 263–275.

Yuan, Y., Zhang, N., Tao, W., Cao, X., He, Y., 2014. Fatty acids as phase change materials: A review. Renewable and Sustainable Energy Reviews, 29, 482–498.

Huang, M. J., Eames, P. C., Norton, B. 2004. Thermal regulation of building-integrated photovoltaics using phase change materials. International Journal of Heat and Mass Transfer, 47(12), 2715-2733.

Mondal, S. 2008. Phase change materials for smart textiles–An overview. Applied Thermal Engineering, 28(11), 1536-1550.

Salunkhe, P. B., Shembekar, P. S., 2012. A review on effect of phase change material encapsulation on the thermal performance of a system. Renewable and Sustainable Energy Reviews, 16(8), 5603-5616.

Karkri, M., Lachheb, M., Nógellová, Z., Boh, B., Sumiga, B., AlMaadeed, M. A., Fethi, A., Krupa, I. 2015. Thermal properties of phase-change materials based on high-density polyethylene filled with micro-encapsulated paraffin wax for thermal energy storage. Energy and Buildings, 88, 144-152.

Sharma, A., Tyagi, V.V., Chen, C.R., Buddhi, D., 2009. Review on thermal energy storage with phase change materials and applications. Renewable and Sustainable Energy Reviews, 13, 318–345.

19

Economic Analysis of Solar Energy Devices

Priyabrata Santra, S. Poonia and R.K. Singh

ICAR-Central Arid Zone Research Institute, Jodhpur, Rajasthan, India

Life cycle cost analysis

Economics of solar energy devices can be calculated through life cycle cost (LCC) analysis. Total life cycle cost of a device is comprised of capital cost, maintenance cost, replacement costs for damagaed components and operational cost. Before adding these above costs, all future costs (C) are converted to present worth considering the relative rate of inflation (i) and discount rate (d).

$$PW = C \times \left[\frac{(1+i)}{(1+d)} \right]^n \tag{1}$$

Where, PW is the present worth of any future cost, i is the relative rate of inflation and d is the discount rate per year and n is the time period in years. Relative rate of inflation accounts for the escalted increase or decrease in prices of a commodity in comparison to general inflation rate. For any commodity, if the price escalation is expected as per the general inflation rate then relative rate of inflation is considered zero. In general case, relative rate of inflation is considered zero. Discount rate accounts the real value of money in future and in most of the economies of the world it is about 8-12%, and therefore 10% is considered in most analyis. The discount rate also refers to the interest rate used in discounted cash flow (DCF) analysis to determine the present value of future cash flows. The discount rate in DCF analysis takes into account not just the time value of money, but also the risk or uncertainty of future cash flows; the greater the uncertainty of future cash flows, the higher the discount rate. A discount rate of 10 % per year would mean that in real terms it makes no difference to a farmer whether he has ₹ 100 now or ₹ 110 in one year's time.

Conversely, a cost of ₹ 110 one year from now has a present worth of ₹ 100. For a future single cost in nth year, the present worth of that cost is calculated using Eq (1). However, for future multiple payments, costs are to be converted to present worth for each year and then needs to be cumulated. For calculation of annualized life cycle cost (ALCC) of solar devices, annuity factor (AF) needs to be calculated for a period of life cycle of the device as follows:

$$AF=\frac{\frac{1+i}{1+d}-1}{\frac{1+i}{1+d}\left[\left(\frac{1+i}{1+d}\right)^{n}-1\right]} \tag{2}$$

$$ALCC=LCC\times AF \tag{3}$$

In case of Solar PV devices, since the PV module works for 25 years after its installment, the life cycle of solar PV devices is conidered 25 years. In case of solar thermal device, the life cycle may be smaller than solar PV devices and accordingly AF to be calculated.

Payback period

The payback period is the time required for the amount invested in an asset to be repaid by the net cash outflow generated by the asset. The payback period is expressed in years and fractions of years. For example, if a farmer invests ₹ 15,000 for a solar device, and it produces cash flow of ₹ 5,000 per year, then the payback period is 3.0 years (₹ 15,000 initial investment / ₹ 5,000 annual payback). An investment with a shorter payback period is considered to be better, since the farmer's initial outlay is at risk for a shorter period of time. The calculation used to derive the payback period is called the payback method. In this method, starting from investment year net cash flow for each year is calculated by subtracting the cash outflow for that year from cash inflow for that specified year. Then cumulative cash flow is calculated by adding the individual net cash flow for each year. The year at which cumulative cash flow becomes positive number, is the payback year. In the payback analysis, ALCC as shown in Eqs 1-3, can be considered as cash outflow for a year. The cash inflow is calculated by adding the income generated by the farmer and the cost of energy saved by using the solar device.

The payback period (*N*) can also be calculated by using mathematical formula as follows (Nahar, 2001):

$$N = \frac{\log\left[\frac{E-M}{a-b}\right] - \log\left[\frac{E-M}{a-b} - C\right]}{\log\left[\frac{1+a}{1+b}\right]} \quad (4)$$

Where, E is the energy savings per year (₹); M is maintenance cost per annum (₹) and C is cost of the device (₹), a is the Compound interest rate per annum and b is inflation rate in energy and maintenance per annum.

LCC analysis of solar PV pumping system

Capital cost of SPV pumping system

The capital cost of a solar PV pumping system is contributed by PV panel, AC pump, inverter, mounting structure, accessories/cables and miscellaneous cost including profits of the seller of these capital items. Life cycle costs for a 3 HP and 5 HP solar PV pumping systems have been presented here since these are commonly available in market and also are under the Govt. subsidy scheme. PV panels are the major contributor to capital cost of a solar PV pumping system and the size of it is governed by the pump capacity or wattage, inverter efficiency and available solar radiation for a location. Efficiency of commonly available inverter is about 85-95% and thus 0.9 is considered here. In order to run the solar PV pumping system throughout the year and for a minimum period of 6 hours a day from morning 10:00 am to afternoon 4:00 pm, the lowest amount of radiation available during this period in a year for a particular location plays the key role to determine the size of PV panel. Generally, a PV panel with a rated capacity of, for example, 100 W_p will generate 100 W power if available solar radiation is 1000 W m^{-2} and the ambient temperature is 25°C. Otherwise, we can say that the PV panel output is proportional to the available solar radiation, which means that if available solar radiation is 700 W m^{-2} during morning time, output from a 100 W_p capacity PV panel will be 70 W. From the available solar radiation data from Jodhpur, Rajasthan it has been observed that lowest amount of radiation available on a tilted surface during morning or afternoon time is 700 W m^{-2} in a year, a factor of 0.7 is therefore considered to determine the size of PV panel in a solar PV pumping system. Likewise, the PV panel size of a 3 HP and 5 HP solar PV pumping system was calculated by dividing the pump wattage by inverter efficiency factor of 0.9 and PV panel factor of 0.7, which were about 3500 W_p and 5900 W_p, respectively. Considering the present day PV panel price of about ₹ 40/W_p, the cost for PV panels in 3 HP and 5 HP solar PV pumping system is about ₹ 1,40,000 and ₹ 2,36,000, respectively which leads to a total capital costs of ₹ 2,92,250 and ₹ 4,36,800 after adding other capital costs.

Maintenance, replacement and operational cost of SPV pumping system

Maintenance cost for a solar PV pumping system is less as compared to others and is considered 1% of the capital cost per year. The maintenance cost is recurring in nature and need to be spent throughout the year and thus the cumulated discount factor for total life of 25 years was calculated, which is 9.08 corresponding to a discount rate of 10% and relative rate of inflation of zero. Thus, the maintenance cost of a 3 HP solar pumping system throughout its life was found ₹ 26,536 whereas for 5 HP system it was ₹ 39,661. It was considered that the AC pump of a solar PV pumping system needs to be replaced after 8 years of its operation and thus needs replacement for two times in its life cycle; one is at 8th year and another is at 16th year. Therefore, the discount factor for 8th year and 16th year were calculated using the Eq (1) and considering the discount rate of 10% and relative rate of inflation of zero, which were 0.47 and 0.22, respectively. The present worth of these two future replacements of AC pump was calculated for both 3 HP and 5 HP systems, which were about ₹ 21,700 and ₹ 24,150, respectively. Similarly, the replacement cost for inverter at 10th and 12th year of its life cycle was calculated for both the system, which were ₹ 10,800 and ₹ 13,500, respectively for 3 HP and 5 HP systems (Table 19.1).

Life cycle cost of SPV pumping system

Component costs and total life cycle cost of 3 HP and 5 HP solar PV pumping system is given in Table 19.1. Total life cycle cost for 3 HP solar PV pumping system was found ₹ 3,51,286 whereas it was ₹ 5,14,111 for 5 HP system. For comparison with other pumping systems, these life cycle costs were converted to annualized life cycle cost (ALCC), by dividing the LCC with cumulated discount factor for 25 years (9.08). The ALCC for 3 HP and 5 HP solar PV pumping systems were ₹ 38,688 per year and ₹ 56,620 per year, respectively.

Table 19.1: Life cycle cost of solar PV pumping system for irrigation

Sr.No.	Parameters	Cost of solar PV pumping system (₹)	
		3 HP system	5 HP system
1.	Life cycle	25 years	25 years
2.	PV panel cost (₹ 40/W_p)	1,40,000	2,36,000
	AC pump cost	30,000	35,000
	Inverter cost	20,000	25,000
	Mounting structure	25,000	30,000
	Cables and accessories	10,000	10,000
	Miscellaneous cost (30% of total cost)	67,500	1,00,800

	Total capital cost	2,92,250	4,36,800
3.	Lifetime maintenance cost (1% of the capital cost)	26,536	39,661
4.	Replacement cost of AC pump (at 8th year and 16th year)	21,700	24,150
5.	Replacement cost of inverter (at 10th year and 20th year)	10,800	13,500
6.	Total life cycle cost	3,51,286	5,14,111
7.	Annualized life cycle cost	₹ 38,688 y^{-1}	₹ 56,620 y^{-1}

Payback period of solar PV pump

To compute the payback period of solar pumping system, apart from ALCC as mentioned above, net cash inflow also to be calculated. The net cash inflow is mainly contributed by the savings of cost for diesel or electricity which otherwise needs to be spent for operating irrigation pumps. Apart from it, the assurance of irrigating the crops at right time is expected to improve the crop productivity, which may be about 10-15% increment and thus contributes in net cash inflow. However, if we ignore the contribution of cash inflow assuming that there is no crop yield improvement by using solar pumps, the net cash inflow is only contributed by the savings of cost for diesel and electricity. If we consider average use of diesel pump in a year for irrigation is 60 days and 6 h day^{-1} the cost of diesel per year for a 3HP diesel pumping set will be ₹ 13,537. Other factors considered in the above calculation are (i) diesel price @ ₹ 56.82 l^{-1} as on 1st September 2016 at Jaipur, (ii) energy value of diesel is 10.5 kWh l^{-1} (iii) diesel pump efficiency is 30-35% and (iv) energy generation by diesel pump is 3.4 kWh per litre. Thus a farmer can save ₹ 13,537/- per year if he/she installs solar pump by replacing a diesel pump, which can be considered as cash inflow per year. Here, it is assumed that cash inflow per year is same throughput its life cycle although it may not be in reality because of price variations of diesels and the discounted cash value in future. The annual cash outflow for a 3 HP solar pump system is ₹ 38,688 as mentioned in Table 19.1. Therefore the payback period of a 3 HP solar pump will be 2.9 years. Similarly, if the solar pump replaces grid connected electric pumps, it saves electricity charge of ₹ 4050/- in a year considering the similar duration of its use as mentioned above and electricity tariff of ₹ 5 kWh^{-1} and thus the payback period will be 9.6 years. Therefore, it is seen that replacing diesel operated pumps by solar pump is highly beneficial to a farmer.

Summary

Life cycle cost analysis of solar energy devices is discussed here. In this analysis, all future costs are converted to present value considering the relative rate of inflation and discount rate or interest rate. Total life cycle cost is multiplied with annunity factor to calculate annualized life cycle cost, which may be considered as cash outflow in each year. The payback period can also be calculated by computing net cash flow in a year. As an example, LCC and payback period of solar PV pumping system is discussed. The ALCC for a 3 HP solar pump is calculated as ₹38,688 whereas the payback period is 2.9 years if replaces diesel pump and 9.6 years if replaces grid connected electric pumps.

Reference

Nahar, N.M., 2001. Design, development and testing of a double reflector hot box solar cooker with a transparent insulation material. Renewable Energy, 23: 167–179.

20

Advances in Solar Photovoltaic Technology: Monocrystalline to Flexible Solar Panels

P.C. Pande

Former Principal Scientist and Head of Division
ICAR-Central Arid Zone Research Institute, Jodhpur, Rajasthan, India

Introduction

Solar energy could be converted to electricity both through thermal route and photovoltaic cells but the latter is more advantageous as it has no moving part, it doesn't require water, a rare commodity specifically in arid regions. Moreover, the maintenance of PV systems is easy, it can be installed right at the place of utility, the system is modular and these are reliable having life more than twenty years and of course eco-friendly.

In photovoltaic cells, generally known as solar cells, charge carriers are generated by the incoming irradiance (energy greater than band gap) and these are separated and swept away by an internal electric field, which is created by either making metal semiconductors contact (Schottky barrier) or p-n junction or MIS (Metal Insulator Semiconductor) or SIS (Semiconductor Insulator Semiconductor) devices. The height of potential barrier at the junction determines the voltage and the flowing charge carriers contribute to the current when connected externally. The choice of material depends on its band gap, absorption coefficient, diffusion length, mobility of charge carriers, ease in contact formation, type of junction with thermal and electronic compatibility, availability of material, toxicity and several other fabricational opto-electronic and practical aspects. PV cells based on single crystal and polycrystalline silicon, amorphous silicon, CdS, CdTe, CuInGaSe, InP, GaAs etc. have been much studied. PV modules of efficiency above 16% on silicon and 10% on CdTe, CuInGaSe and

8-9 % on amorphous silicon are commercially available. Dye sensitized TiO_2 devices have also been receiving more attention with coloured panels for windows. Organic semiconductors are explored and more recently perovskite solar cell has been studied. Flexible solar cell panels based on thin film devices are in use. The applications have gone from solar lighting to operation of pumps, small and medium size equipment, stand alone and grid connected solar power plants, roof top and building integrated PV and now even energizing cars and aero planes. With technological advancement and enhanced production (233 GW in 2015) the cost of commercially available PV modules, which used to be high at one time has been brought down from 30 \$/Wp in seventies to about \$0.5/Wp at present. Here Wp means the maximum power output when sun is at zenith at sea level i.e. the radiation values are 1000 W/m^2 and ambient temperature is 25^{0}C. In fact with this cost the solar PV electricity is approaching grid parity. If utilized aptly, it is most appropriate technology for remote rural applications. In this connection, it becomes important to know about the chronology, fabrication, the recent trends and status of these various forms of solar cells.

Historical Background

The concept of the photovoltaic cell is an old one. As long ago as 1839 (Unit of Radioactivity), a French scientist Becquerel discovered that a photovoltage resulted from the action of light on an electrode in an electrolyte solution. Subsequently some pioneering work was performed on selenium and cuprous oxide photovoltaic cell. This work eventually resulted in the photovoltaic exposure meter, which was used up to 1950's.

The modern history of solar cell originates in 1954 when Chapin and coworkers of Bell Laboratories, USA, reported a 6% efficient single crystal silicon solar cell. In the same year 6% Reynolds and coworkers fabricated efficient CdS solar cell and in 1956 Jenny and colleagues in USA also demonstrated 4% GaAs solar cell. At that time, the importance of photovoltaic cell was mainly in the space programmes and the prime focus of research was to increase the efficiency and reliability of the cells. Particular attention was paid to solving the problem of radiation damage to the cells in the space. CdS solar cells were considered to be more radiation resistant and their use in thin film form offered the prospects of enhancing the power to weight ratio. However, these CdS cells failed to reach a sufficient high efficiency to compete with single crystal Si solar cell and GaAs based device. In view of the interest in utilizing photovoltaic cells for terrestrial purposes, reducing their cost has become the primary aim. To achieve this, improvements in the technology and an increased scale of production of silicon-based solar cells were given priority. The production mainly includes modules based on single crystal silicon cells and polycrystalline silicon based photovoltaic cells. While attempts are made to bring down the cost further through improved technologies, more attention is given to promising thin film solar cells. The thin film solar cells require much less material and energy for their fabrication than single crystal cells. Several materials viz. amorphous silicon, CdS, CdTe, $CuInGaSe_2$, InP, GaAs etc. have been investigated. Some new materials, organic semiconductors were also explored and off late studies on quantum dot solar cells were conducted. With these efforts it is expected that the cost could be reduced substantially.

Choice of Material

The essential features of a solar cell are an absorber generator material in which mobile carriers are created by the absorbed solar energy and a built-in voltage, which allows the generated carriers to be collected from region in which they are produced and converted to majority carriers. The purpose of the collector converter is to prevent the back flow of carriers. The absorber region controls the magnitude of the current that is generated and the height of the potential barrier determines the voltages that the cell can produce. Obviously to produce more photo current the band gap of the semi-conductor should be small while on the other hand, in order to obtain high open circuit voltage a large value of band gap is preferable. When these two factors are matched with the solar spectrum, it is found that the optimum value of band gap of the material comes to 1.44 eV i.e. for CdTe. However, in the range of 1.0 eV to 1.7 eV efficiencies above 20% can be achieved.

The generator material should have a high value of absorption coefficient to ensure capture of all available photons. When the spectral density is convoluted with the absorption coefficient of the materials suitable for photovoltaic cells, one finds that CdTe will absorb 90 percent of the sunlight above its energy gap with a thickness of only 0.4 micron, while silicon needs 100 micron thickness for the same purpose.

The generated carriers must be mobile and must continue in their separated state for a time that is long compared with the time they require to travel to the localized charge separating in homogeneity. This process is characterized in terms of the carrier diffusion length, which may be viewed as the distance over which the photo-generated carrier density decreases by e^{-1} as the carriers move by diffusion. The diffusion length increases as the square root of the product of the recombination lifetime and mobility of the carriers. The recombination lifetime depends on the capture cross section and density of defects. The mobility of the carriers, in turn, depends on the scattering mechanism present. All of these parameters depend on temperature, the impurity concentration, crystallinity, crystal orientation and type of defect. It is estimated that to transport 90 per cent of the generated minority carriers to the junction, the diffusion length should be twice the absorber film thickness.

The internal electric field within the semiconductor is created by an electronic inhomogeneity, which can be achieved either by providing a metal semiconductor contact (Schottky barrier) or by forming a p-n junction between two regions of a semiconductor (homo junction) or a junction between two different semiconductors (hetero junction). There are other structures also like metal insulator semiconductor (MIS), semiconductor insulator semiconductor (SIS) devices.

Homo junctions have the advantage that their theory has been studied in details because of their applications in rectifiers and transistors. Some metallurgical and electronic problems such as matching of thermal expansion coefficients, lattice constants and electronic affinities do not arise in this junction. However, appreciable losses of the mobile carrier take place due to front surface recombination. Since only a few materials can be doped both p and n-type, useful homo-junctions are limited to single crystal silicon, amorphous and polycrystalline silicon, gallium arsenide and cadmium telluride.

A hetero junction solar cell consists of a small band gap semiconductor (absorber generator) in which optical absorption takes place and a large band gap material (collector converter) that acts as a window for the junction. If the two semiconductors have the same type of conductivity the hetero junction is called isotype, otherwise it is known as anisotype. Such hetero junctions can be classified as abrupt or graded according to the distance in which the transition from one material to the other is completed near the interface. The primary advantage of a hetero junction structure is that it allows various materials, which cannot be, doped both p and n-type but have other outstanding features. Moreover, the junction can be operated in front wall or back wall modes. A suitable material is chosen for the absorber generator, which has all characteristics, discussed in the preceding section. For the collector converter the band gap should be as large as possible while maintaining a low series resistance. The other important factors are the lattice constants and electron affinities of the two semiconductors. A high density of interface states is introduced by the lattice mismatch between the two semiconductors, and band discontinuities develop because of the difference in electron affinities. With the proper choice of a ternary compound as collector converter, these drawbacks can be overcome and a hetero junction can exhibit the optimum properties of a homo junction without the problem of front surface recombination loss.

The positive aspects of homo junctions and hetero junctions are combined in heteroface structures in which the free surface of a homo junction is replaced with a large band gap window materials so that the original free surface recombination velocity is replaced by an interface recombination velocity which is several orders of magnitude smaller. The most common heteroface structure is the $Al_x Ga_{1-x} As$/ GaAs solar cell.

The Schottky barrier has the advantage of ease of preparation since it does not require diffusion processes to be carried out at the elevated temperature, and a simple blocking contact is formed. Large thermionic emission currents that reduce the open circuit voltage usually limit performance of these devices. Metal Insulator Semiconductor (MIS) and Semiconductor Insulator Semiconductor (SIS) junction have received much attention. These are equivalent to Schottky barrier and hetero junction systems respectively, with the addition of a thin layer of insulator usually an oxide at the interface to reduce the forward current.

A recent concept to increase the efficiency is that of Cascade/Tandem solar cells. Since a single semiconductor utilizes only a limited portion of the incident solar spectrum and the open circuit voltage of the device is limited by the band gap of the semi conductor, these two effects lead to high internal losses in conventional cells. If two or more solar cells having different semi-conductor materials with suitably separated band gap values are made one behind the other, such that the largest gap material faces the incident radiation first, the high energy photons are absorbed by the first materials, and the rest of the solar spectrum fall on the second solar cell which absorbs the higher energy portion of the transmitted radiations while the remainder passes to the third one. This selective absorption process continues down to the cell with the lowest energy gap. Alternatively the incident photons are split into spectral parts by an optical filter, and each part is directed towards a separate cell, which is designed to match with a specific part of the spectrum.

The selection of material and type of junction also requires an investigation of the availability of material cost, material toxicity, cell stability and lifetime. It should also be possible to form low resistance electrical contacts to both n and p type materials. Since all the semiconductor materials used for solar cell have high refractive indexes, 25-35% of the incident radiation is reflected from the planar surface. Anti-reflection coatings are used to minimize these losses. Generally the surface is textured and then a layer of AR coating is applied. SiO_2 has successfully been used on Si solar cells. A layer of TiO_2 is used in GaAs devices. Finally, the cell should be well encapsulated to protect the solar cell from the environment.

From these considerations, it is clear that only a few limited materials can be used for solar cells. Besides, silicon or gallium arsenide other materials which have been explored are CdS, CdSe, CdTe, $CuInSe_2$, InP, GaAlAs, amorphous silicon etc.

Fabrication of Solar Cells

Single crystal silicon

Everybody is aware of availability of silicon in the form of sand. It is second largest available element in the world. However, the extraction of pure silicon is quite tedious process. First of all metallurgical grade silicon (98% pure) is obtained by reducing quartz in arc furnaces using charcoal. This is further purified to ppb level. For this MG silicon is first melted and allowed to react with gaseous HCl in presence of catalyst like Cu in fluid bed reactor to form purified $SiHCl_3$ which is further reduced to polycrystalline Si in reduction furnaces using Siemens process. The cost of this material is reflected in the price of solar cell. Subsequently single crystal boules are grown by Czochralski technique. In this process an oriented single crystal seed is first dipped in molten silicon and the slowly withdrawn with sophisticated puller which rotating silicon crucible containing molten silicon and seed. The power is reduced slowly till the crystal is obtained. Typically Czochralski grown single crystal p type silicon has a resistivity of 1-10 ohm cm. Subsequently 300-400 micron thick and 10-12 cm diameter or even larger wafers are obtained by cutting the boule. There is a great improvement in sawing technology. With wire sawing there is now much less wastage. Solar cell formation involves further several processes which include removal of mechanically damaged surfaces by chemical etching process, back surface field formation by doping boron, junction formation through diffusion of phosphorus, providing ohmic contacts through grid and back surface metallization, scribing and etching the edges to avoid leakages, providing antireflection coating, etc. This simply indicates the number of processes required to fabricate the cell. It is worthwhile to mention about the improvement in the efficiency by Martin Green and co workers of Australia who provided textured surface and laser grooving for making buried contact in order to reduce surface recombination. These ideas of Passivated Emitter Solar Cells (PESC) and Back Point Contact solar cells made it possible to break 20% efficiency barrier and efficiency above 25 % on small area cell has been achieved.

Polycrystalline Silicon

The other form of silicon is polycrystalline silicon, which is made up of many grains of single crystal silicon. The polycrystalline silicon is easy to grow through directional solidification in which molten silicon is cooled slowly along one direction. Commercially casting processes for making large grains (greater than cm size) is used to grow polycrystalline silicon, which is simple and cheaper and has high through put. However, grain boundaries act as sink for generated charge carriers. Therefore, grain boundary passivation is required to achieve efficient solar cells. Reports are available for cells up to 21% with multicrystalline material by using Plasma-enhanced chemical vapour deposition (PECVD) PECVD SiO / SiN deposition and forming gas anneal for defect passivation and Al treatment for defect and impurity gettering.

Silicon films on foreign substrate have been grown by methods like edge-defined growth (EFG), dendrite web growth, ribbon to ribbon growth etc. Efforts are made to develop Silicon sheet with a cheaper way and this may find a lot of promises. Simultaneously, since solar cell does not require as pure Si as required for electronic purposes solar grade silicon is being used in an attempt to reduce the cost. Meanwhile systems based on available PV modules have been developed which include both stand alone as well as integrated systems.

Thin Film Solar cells

The thin film solar cells require less material and energy for their fabrication than single crystal cells. The module fabrication is also simple because all cells can be interconnected monolithically with ease compared to single crystal where each cell is placed and then connected externally. Several materials viz. amorphous silicon, CdS, CdTe, $CuInSe_2$, InP, GaAs etc. have been investigated. CdS-Cu_x S devices, having high potential for low cost cells, were studied in detail. The device came very close to commercial production by Photon Power Inc. in USA and Nukem, GmbH but degradation beset the progress. Subsequently more stable CdTe and $CuInSe_2$ as absorber replaced Cu_x S.

CdS-CdTe devices

CdTe has been considered to be attractive material particularly because it has optimum energy band gap of 1.45 eV and large absorption coefficient with theoretical efficiency of the order of 29%. Generally, the structure of the device is glass/SnOx/CdS/CdTe/ohmic contact. Various methods have been used to deposit CdS and CdTe layers and successful junctions have been demonstrated by using methods viz. thermal evaporation, electroplating, screen printing and close spaced sublimation. Some work was done on electrophoretical deposition of CdTe and then laser induced recrystallization for developing a low cost technique for CdS-CdTe solar cells. A post barrier heat treatment with $CdCl_2$ in presence of air at 400°C for a short duration is administered to attain higher efficiency. The enhancement in the device parameters with this heat treatment is attributed to increase in the grain size and removal of deep traps.

CdS-CdTe solar cells contribute to about 5% of total world solar cell production and potential to produce cheapest thin film solar cells. Such devices have been prepared on flexible substrates and have advantage to put these on curved surfaced. The apprehension of cadmium safety has been studied in great detail and the devices conform to the international standards. The other factor is the requirement of rare Tellurium, although projections indicate sufficient availability even with several GW productions.

CdS-Cu(InGa)Se_2 Devices

$CuInSe_2$ (CIS) is a direct band gap chalcopyrite semiconductor having a band gap of 1 eV and relatively higher absorption coefficient, which is suitable for producing CdS-$CuInSe_2$ devices with theoretical efficiency of 28.6%. Several processes which have been used for the deposition of CIS and CIGS include multiple sources evaporation, selenization of evaporated or sputtered Cu/In precursors in H_2Se or elemental Se atmosphere, electrolytic deposition, silk screen printing, sputtering, laser ablation etc. The basic structure is glass/Mo/CIS/CdS/ZnO/grid. The molybdenum back contact is deposited by sputtering or electron beam evaporation. After depositing CIS or CInGaS layer as absorber, a thin buffer layer of CdS of about 500 nm is prepared in a chemical bath. The front contact consists of ZnO, which is fabricated by RF sputtering or by a CVD process. Efficient devices have been produced by depositing the CIS layer either by three-source evaporation process through the use of effusion source or by selenization of Cu-In deposited on the substrate and subsequently heated in furnace at higher temperature in H_2Se ambient that selenizes the film. However, efforts have been made on the development of low cost techniques like electrolytic deposition and silkscreen printing. Although, efficient devices have been shown by these techniques, there are some difficulties in producing the proper stoichiometry. In the initial devices a thick layer of CdS was used as a window. Then subsequently, thin films grown by chemical bath method have resulted in achieving higher efficiencies. In this device the thickness of CdS is only 5000 nm which makes it possible to achieve higher Jsc. ZnO, is both transparent electrode and acts as partially reflecting coating.

The major break through came when the devices formed on soda glass containing Na provided excellent results. The role of Na needs to be studied in detail. Various groups have attained efficiencies over 16 per cent. Recently a cell made of copper indium gallium di selenide set a world record of 21.7 per cent efficiency. R. Noufi and his team at NREL, USA, have developed the cell. Commercially CIGS modules of 5, 10, 20 and 40 W are available. Siemens Solar Industries, USA has taken a lead. Global Solar, USA has planned 10 MWp flexible solar cells on polymer foils. The modules based on CIGS cells have been reported to be extremely stable. The cost of the PV modules has not yet been reduced as expected which may be due to poor through put and small-scale manufacturing.

Amorphous Si :H Based Devices

The amorphous silicon cells are based on hydrogenated amorphous silicon (a-Si:H) or fluorinated material, a glassy semiconductor material having approximate band gap of 1.6 eV and high absorption coefficient in order to absorb sufficient incident radiation with in 0.5 micron and thus allowing a thin layer for the fabrication of the device compared to few hundred times thick single or polycrystalline silicon. In fact amorphous silicon has several dangling bonds, which impede the free passage of charge carriers. By incorporating hydrogen atoms or fluorine the number of dangling bonds is reduced giving freedom of movement of electron and holes. Even then, the mobility of charge carrier is very poor compared to crystalline silicon.

Hydrogenated amorphous silicon is deposited by decomposing silane gas. As mentioned earlier, purified silane is reduced to high purity silicon before giving single crystal silicon and thus not only it reduces the requirement of material it reduces the steps required in processing single crystal. There are several structures available of aSi cells. A typical structure comprises conducting glass, p type a Si: H doped with boron (50-100 A thick), intrinsic layer of undoped aSi: H (0.5 micron) an n type aSi : H doped with Phosphorus (200 A thick). The main active layer is undoped a : Si : H. Although the mobility of charge carrier is low, the existing strong electric field sweeps the generated charge carriers. To collect these a layer of metal is deposited. There are several techniques, which are used to fabricate the cells. This includes Glow discharge, sputtering, and chemical vapour deposition. The chemical vapour deposition (CVD) are operated either at a high or a low gas pressure. The gasses react and are deposited on the substrate. Efforts are afoot to fabricate a SiGe : H alloys with band gaps between 1.4 & 1.5 eV and using these for cascading the devices of aSi : H and aSiGe : H cells to obtain higher efficiency. So far the biggest problem has been the degradation of the cell. With time and exposure to light the efficiency goes down and typically 6% efficient module are produced. These are explained by Staebler Wronsky effect and accordingly the hydrogen bonds get weaken and so more disorder state is formed resulting in more traps. United Solar Systems Corp. (USA) has produced 14.6% efficient solar cell module based on aSiGe:H in a unique triple cell design with 13% stable efficiency. The world's highest stabilized efficiencies of 9.5% for an aSi/a-SiGe superstrate submodule (1200 sq.cm) was achieved by combining the cell design technique with other technologies. The cost of aSi solar cells are comparable to single crystal devices due to low stable efficiencies. But these cells can be fabricated on flexible substrate and can be put easily on curved surfaces, windows and facades. With more efficient stable devices, the cost could be reduced further to be more competitive in the market. The other advantage is that these devices can sustain higher temperatures. Recently microcrystalline Si : H tandem devices with amorphous and micro crystalline materials have provided promising results.

Nanocrystalline and Other PV Devices

A photo electrochemical solar cell based on thin film comprising nano particles of TiO_2 and sensitized by a Ru- complex dye has been reported to have potential to produce low cost solar cells. These Dye Sensitized Solar Cells (DSSC) known as Graetzel solar cell with light absorption for carrier transport has also come up with stabilization in the efficiency and reduction in the degradation. In these devices electron transfer should be faster than recombination. Sensitization has to sustain 100 million cycles for 20 year cell operation. These devices are considered suitable for power window applications due to availability of modules in different colours, even attractive panels with flowers makes the availability of different designs for windows. With these devices sunglasses cum charger is a commercial option. Bifacial dye sensitized solar cells, if stabilized, will be excellent for indoor applications. The challenges are to attain stability at higher temperatures and to enhance the number of cycles in dye sensitized solar cells.

Recently more work is carried out on perovskite structures compound such as methyl ammonium lead halides which are cheap to produce and simple to manufacture. A perovskite solar cell is a type of solar cell which includes, most commonly a hybrid organic inorganic lead or tin halide based material, as the light harvesting layer. PV devices based on a blend of organic polymers and an assortment of two types of nanorods-clusters of Cadmium Selenide molecules (7-60 nanometers in diameter) have been reported to have high potential to produce low cost solar cells. Work on nano particles based devices on silicon have been undertaken by Martin Green group in Australia. Organic semiconductors based devices, inclusion of carbon nano rods are some area of investigations. Device based on quantum dots and cascading for getting 60 % efficient devices have high hopes.

Status

The best efficiency on single crystal silicon has been above 25% while that of module 16-18%, polycrystalline 21% and on module 15-16%, amorphous silicon thin film devices 14%, module 8-9%, CIGS and CdTe 21.4 and 22.1% and on module about 10%. The concentrated cells on GaAs have shown an efficiency of 35%. The challenges in thin film solar cells are further reduction in thickness, development of simplified process at lower temperature with proper stochiometry at larger scale with alternative nano materials and technologies.

The annual global production of PV has increased from 3 MW of electricity in 1980 to a capacity of 233 GW in 2015. The cost of solar cells has come down from \$35/Wp in 1970s to less than \$0.5/Wp. Efforts are afoot all over the world to bring down the cost further with novel approaches. Simultaneously appropriate PV systems have been developed for different domestic, agricultural and rural applications. Under National Solar Misssion 100,000 MW solar power plants are envisaged by 2022. Several solar pumps have been installed and solar lighting systems distributed for remote areas. Building Integrated Photovoltaics (BIPV) is rage. PV duster, winnower, dryer, PV enclosures are in operation at CAZRI. It is to be seen how it can be used further in agriculture.

Suggested Readings

Aberle, A.G. 2009. Thin Film Solar Cells, Thin Solid Films 517: 4706-4710.

Becquerel A.E. 1839. Memoire sur les effects electriques produits sous influence. des rayons solaires. Comptes Rendus 9: 561-567

Chu, T. L. and Chu, S. S. 1995. Solid State Electronics 38(3): 533.

Chopra, K. L. and Das. S. R. 1983. Thin Film Solar Cells (Plenum Press) New York.

Duke, S., Miles, R. W., Pande, P. C., Carter, M. J. and Hill, R. 1996. Characterization of in-situ thermally evaporated CdS/CdTe thin film solar cells with Ni-P back contacts. J. Crystal Growth, 159 916-919.

Fahrenbruch, A. L. and Bube, R. H. 1983. Fundamentals of Solar Cells Photovoltaic Solar Energy Conversion (Academic Press, New York).

Graetzel, M. 2003. Dyesesitized solar cells. Journal of Photochemistry and Photobiology C: Photochem Rev. 4: 145-153.

Green, M.A., Wenham, S.R., Honsberg, C.B. and Hogg. D. 1994. Transfer of buried contact cell laboratory sequences into commercial production. Solar Energy Materials and Solar Cells 34(1-4): 83-89.

M.A., Emergy, K., Hishikawa, Y., Warta, W., Dunlop, E.D. 2017. Solar cell efficiency tables progress in photovoltaics: Research and Application 25: 3-13.

Guha, S. and Yang, J. 2009. Advances in amorphous and nano crystalline silicon alloy solar cells and modules. Presented in 18th International Photovoltaic Science and Engineering Conference (PVSEC 18) at IACS, Kollkata, India

Jackson, P., Hariskos, D., Wuerz, R., Kiowski, O., Bauer, A., Friedlmeier, T.M. and Powalla, M. 2015. Properties of Cu (In, Ge) Se solar cells with new record efficiencies to 21.7% Physica Status Solidi 9(1): 28-31.

Pande, P.C. 1984. A study on various forms of CdS solar cells. Ph.D. Thesis. University of Durham, U.K.

Pande, P.C., Russell, G. J. and Woods, J. 1984. The properties of electrophoretically deposited layers of CdS. Thin Solid Films, Elsevier 121: 85-94

Pande, P.C., Bocking, S., Miles, R. W., Carter, M. J., Latimer, J.D. and Hill, R. 1996. Recrystallization of electrophoretically deposited CdTe films. Journal of Crystal Growth 159: 930-934.

Ray, Swati. 2009. Nanocrystalline silicon based thin film solar cells. NACORE. MacMillan India. pp. 267-272.

Rohatgi, A, Rintow, A, Das, A and Ramanathan, S. 2009. Road to cost effective solar PV. PVSEC 18, Calcutta, India

Schock, H. W. and Pfisterer, F. 2011. Thin film solar cells: Past Present and Future Renewable Energy World 4(2): 75-87.

Sharma, G.D. 2009. Advances in nano structured organic solar cells. NACORE, pp. 284-297.

Turkestani, M. K. Al, Major, J.D., Trehane, R.E., Proskuryakov, Y.Y. and Durose, K. 2010. Recent research trends in thin film solar cells. Annals of Arid Zone 49 (3&4): 303-10

Ullal, Harin S. 2009. Polycrystalline solar technologies. Presented In PVSEC 18, Calcutta.

21

Phase Change Materials A Sustainable Way of Solar Thermal Energy Storage

Abhishek Anand*, A. Shukla, Atul Sharma

Non-Conventional Energy Laboratory
Rajiv Gandhi Institute of Petroleum Technology, Jais, Amethi, India

Introduction

As the most of the conventional resources are depleting and can last only for about a few 30 to 50 years. Scientists worldwide are looking for alternative energy resources and they have come to an agreement that solar energy can provide much of our energy needs in the future. The major challenge before them is to store this energy. The radiation falling on the earth surface can be tapped and stored in no of ways. However, the most efficient way is done through the latent heat storage through PCM. The PCM has many added advantages i.e. high energy density, isothermal behavior, non-polluting and low cost. Nowadays, it gaining popularity worldwide and now seen as the most efficient way of thermal energy storage. There is a large number of PCM at use i.e. organic, inorganic, eutectic and salt hydrates. Their utilization depends on the user's requirement technological constraints. They also serve the purpose of energy availability at night beside the uninterrupted supply during the day. They are seen as the materials for ensuring the environmental sustainability.

Energy storage methods

The energy can be stored in basically three methods viz. Mechanical, electrical and thermal energy.

Mechanical energy storage: The basic forms of mechanical energy storage are flywheels, compressed air energy storage (CAES), pumped hydropower storage (PHPS). Nowadays carbon nanotubes springs are gaining popularity as the innovative mechanical energy systems.

Electrical energy storage: Electrical energy is stored through the secondary rechargeable batteries, capacitors, supercapacitors, superconducting magnetic energy storage (SMES).

Thermal energy storage: The various forms of thermal energy storage are brick storage heater, cryogenic energy storage, liquid nitrogen engine, eutectic system, ice storage air conditioning, molten salt storage, phase change materials (PCM), solar ponds, steam accumulator

The basic thermal energy storage is done through sensible heat storage (SHS) and latent heat storage (LHS) or sometimes the combination of both. These basic forms of thermal energy storage are listed in Fig 21.1.

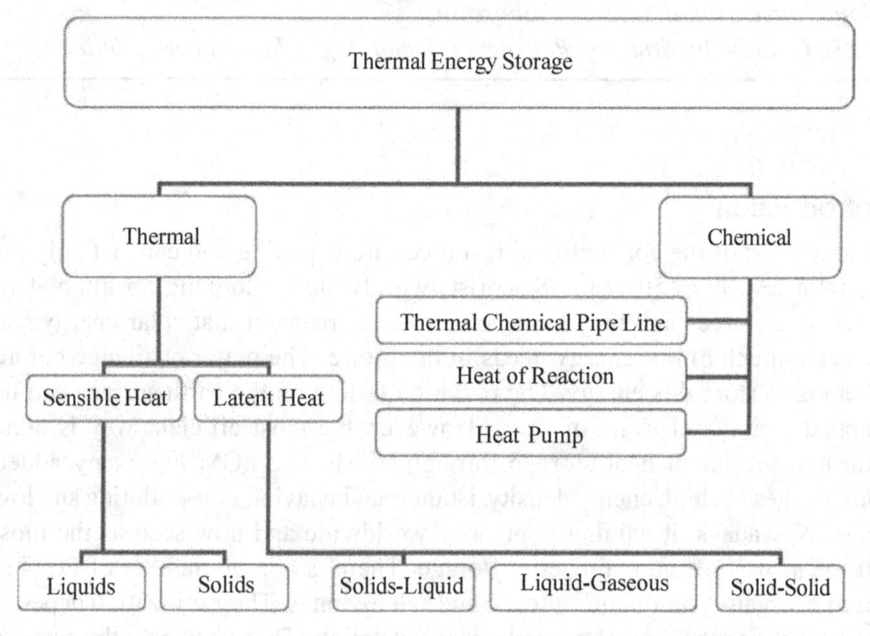

Fig. 21.1: The basic forms of thermal energy storage (Sharma *et al.*, 2009)

Sensible heat storage: In the sensible heat storage, the energy is stored with the change in the temperature of solid or liquid. This type of energy storage depends on the specific heat of the material. The equation governing the sensible heat storage is,

$$Q = \int_{T_i}^{T_f} mc_p dT = mc_p \Delta T$$

where Q is the heat energy stored in (J), m is the mass of the material in (Kg), c_p is Specific heat of the material in (J/Kg K), T_i is the initial temperature in (°C) and T_f is the final temperature in (°C).

Water is one of the best material for the sensible heat storage because it has high specific heat and abundant availability. The major constraint with water is that it can be used only up to 100 °C. For the higher temperature applications, molten salts can be used.

Latent heat storage: Latent heat storage is based on the heat change during the phase change undergone by materials. The heat storage potential of a material undergoing phase change is governed by,

$$Q = \int_{T_i}^{T_m} mc_p dT + a_m \Delta h_m + \int_{T_m}^{T_f} mc_p dT$$

Where a_m is the fraction melted and h_m is the heat of fusion per unit mass (J/Kg) and rest has the meaning discussed above.

In the chemical reaction, energy is stored or released through the breaking and formation of bonds.

Classification of PCMs

The PCM is categorized into three types viz. organic, inorganic and eutectic. A classification of PCMs is given in Fig 21.2.

Organic PCM: The organic PCM is the paraffin and non-paraffin compounds. The paraffin compounds are generally hydrocarbons. The non-paraffin compounds are the fatty acids such as stearic acid, myristic acid, palmitic acid etc.

Inorganic PCM: This class of PCM consists of salt hydrates and metallic. The major advantages of using this class of PCM are they do not degrade with multiple cycling and they do not suffer cooling easily.

Salt Hydrates: They are used extensively for the thermal energy storage because they offer some extra advantages which other classes of PCM fail to offer. The major advantages of using this class of PCM are many. Some of them are: (i) it offers high thermal stability and conductivity, (ii) little or no volume change during phase change, (iii) Non-corrosive and economic availability.

Metallics: Metallics are low melting metals. They are not common candidates for use in the PCM technology because of weight penalties. But they can be promising because of their high conductivity and high heat of fusion per unit volume. The advantages of using them are previously mentioned.

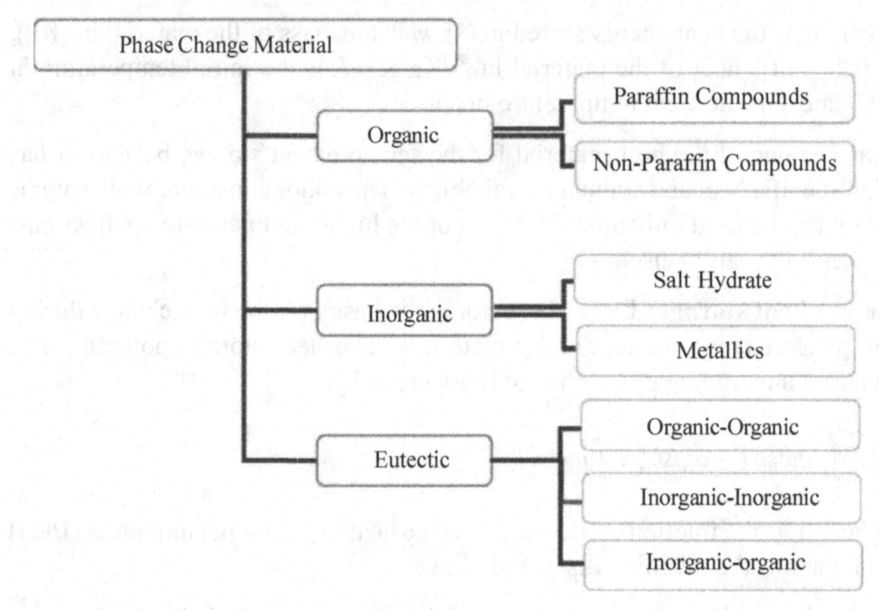

Fig. 21.2: Classification of phase change materials (Sharma et al. 2009)

Eutectics: the eutectics are generally the mixtures of two or more than two components. They must have the homogenous mixture to avoid the separation. The melting point is generally between the two mixtures.

Measurements technique of latent heat of fusion and melting point

The current measurement technique for the melting point and latent heat of fusion is through (i) Differential Scanning calorimeter (DSC) and (ii) Differential Thermal analysis (DTA). In the calorimeter, the sample and the reference material is heated at a constant rate. The temperature difference between them is proportional to the difference in the heat flow between the two materials. For the reference the material commonly used is Alumina (Al_2O_3). The area under the curve gives the latent heat of fusion and the slope give the melting point.

Thermal Energy Storage: A review

Buildings

(Dong *et al.* 2016) presented macro-encapsulated PCM hollow steel ball (HSB), which was embedded with octadecane paraffin used as PCM (Fig. 21.3).The macro-encapsulated PCM-HSB was prepared by incorporation of octadecane into HSBs through vacuum impregnation. Test results showed that the maximum percentage of octa decane carried by HSBs was 80.3% bymass. The macro-encapsulated PCM-HSB has a latent heat storage capacity as high as 200.5 J/g. The indoor thermal performance test revealed that concrete with macro-encapsulate-d octadecane-HSB was capable of reducing the peak indoor air temperature and the fluctuation of indoor temperature. These PCM based balls can be very effective in transferring the heating and cooling loads away from the peak demand times.

Fig. 21.3: Hollow steel ball with hole

In the previous study, author have used the mixture of fatty acid viz. palmitic acid, lauric acid, myristic acid etc and the nano-porous graphite. The necessary processing and the characterization were done. After that, the PCM/Graphite composite was encapsulated in aluminum containers (Fig 21.4). The encapsulated materials were then used in the pilot green building in Shanghai, China. It came out that the thermal conductivity of the PCM is enhanced manifold by the use nano-porous graphite. The payback period for the PCM/Graphite composite was reported to be about 2.5 to 3 years.

Fig. 21.4: PCM Containers (Dong et al. 2016)

(Xu et al. 2013) studied the thermal performance of a prototype room using the shape stabilized PCM. The shape stabilized PCM was made by mixing 70 wt. % of paraffin, 15 wt. % of polyethylene and 15 wt. % by styrene-butadiene-styrene block copolymer (SBS) as the supporting material. The cabin of the prototype was made 3 m in depth, 2 m in width, and 2 min height (Fig 21.5). The south-facing wall consists of 1.6 m×1.5 m double-glazed vacuum window. 100mm thick polystyrene board was used to make roof and wall. The floor was developed with 50mm thick polystyrene insulation layer and 8mm thick shape-stabilized PCM plates. They have shown that the mean indoor temperature of the room with the PCM floor is about 2 °C higher than that without PCM. The indoor temperature swing is also minimized with this experiment. So it was finally concluded that the use of the floor PCM increased the thermal comfort and reduced the space heating in the winter season.

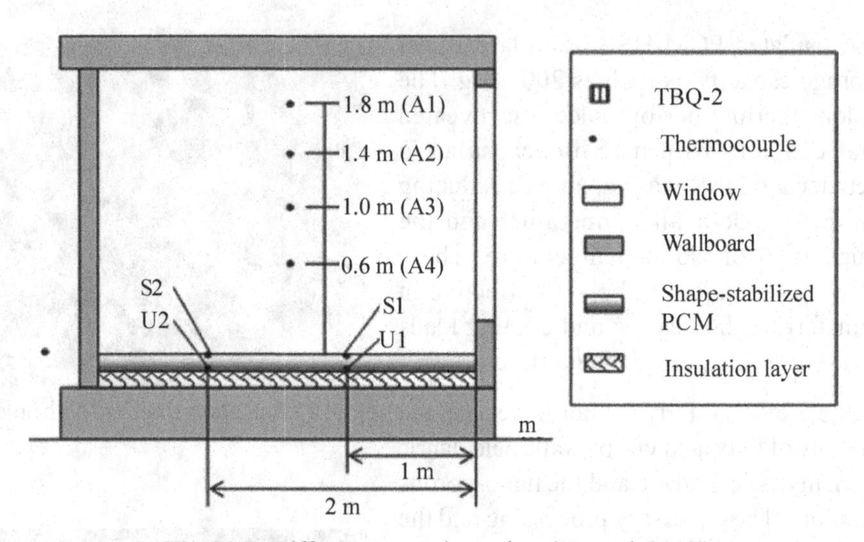

Fig. 21.5: Different measuring points (Xu et al. 2013)

(Sharma et al. 2007) built a prototype of the solar chimney for the ventilation system with the built-in thermal energy storage (Fig. 21.6). They have used Sodium Sulfate deca-hydrate ($Na_2SO_4.10H_2O$) as the PCM. The integration of PCM inside the solar chimney gave the positive results. It came out that the prototype setup could supply constant airflow of $155m^3/h$ in the evening and night if the PCM was completely melted in the day. In the daytime, it was capable of providing $200m^3/h$ airflow.

Fig. 21.6: The prototype of the Solar Chimney with built-in PCM (Sharma et al. 2007)

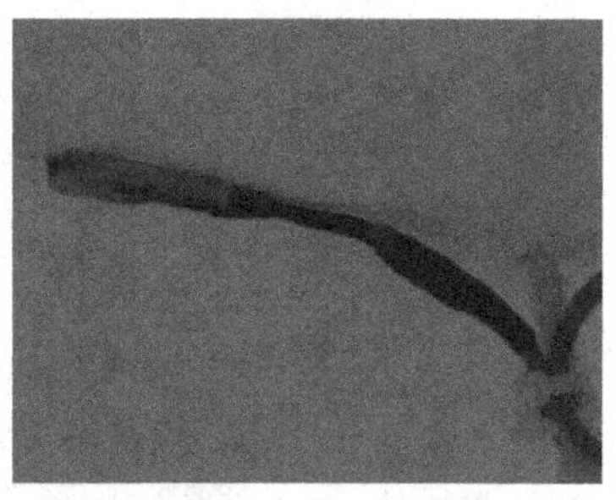

Fig. 21.7: The measurement of the ceiling temperature (Fisch and Kühl, 2004)

(Fisch and Kühl,2004) has studied the effect of PCM as the additive material on the ceiling. For this two-room was monitored, one with PCM and other without PCM. The test was conducted in the summer and it came out that the temperature of the test room was reduced to about 2K by the use of PCM as the additive materials on the ceiling. This could also increase the thermal comfort of the room.

(Rudd 2008) used the coconut oil as the PCM. The test was conducted for the PCM based wallboard. The experiment (Fig. 21.8) showed that the PCM wallboard had the thermal heat storage capacity of 24.2 J/g. The experiment also showed that PCM integration in the building could be a successful proposition for the large-scale thermal energy storage.

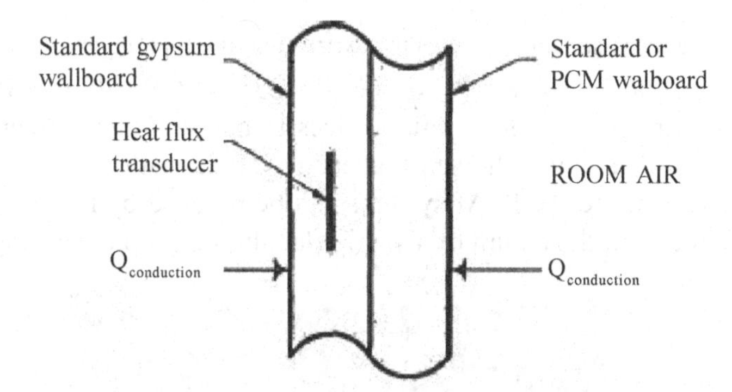

Fig. 21.8: Wallboard energy balance (Rudd 2008)

(Kondo and Ibamoto 2002) studied the effects of the peak cut control of air conditioning system using PCM for ceiling board in the office buildings. He concluded that the PCM ceiling system is effective for the peak-cut control. (Nagano et al. 2006) proposed a new floor supply air conditioning system that incorporates Paraffin wax used as a PCM into building mass thermal storage. A granulated PCM was created by impregnating foamed glass beads with paraffin wax and then applying three coatings of urethane to it. Heat response tests revealed that the temperature stabilization effect during the time when phase change was occurring increased the time constant of the room by a factor of 1.5-2.1 when a PCM packed bed was installed. Results from measurements simulating an air conditioning schedule in office buildings indicated that 89% of daily cooling loads could be stored each night in a system that used a 30 mm thick packed bed of the granular PCMs (Fig 21.9).

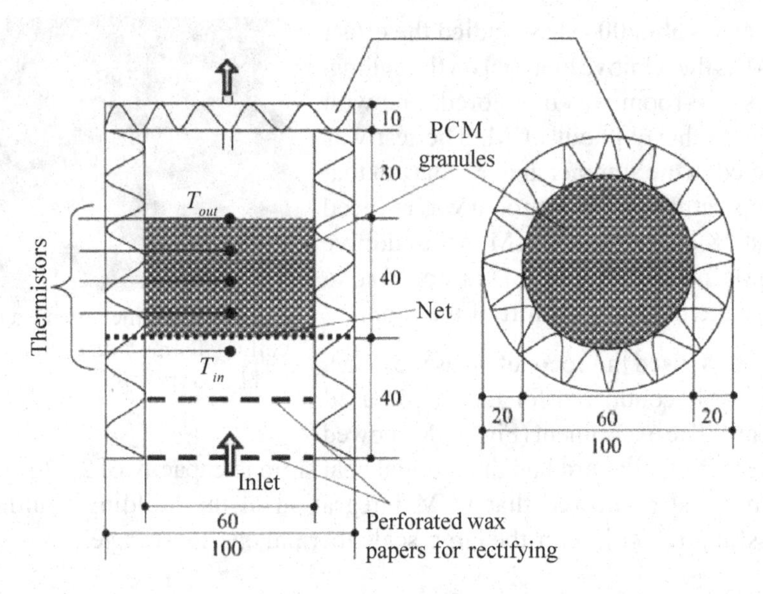

Fig. 21.9: The cross-sectional view of the experimental setup(Nagano et al. 2006)

(Huang et al. 2006) used two PCMs i.e. RT 25 and GR- 40 as a PCM for thermal management of photovoltaic devices to maintain the temperature of the PV system. This study showed that using RT 25 with internal fins, the temperature rise of the PV/PCM systems can be reduced by more than 30°C when compared with the datum of a single flat aluminum plate during phase-

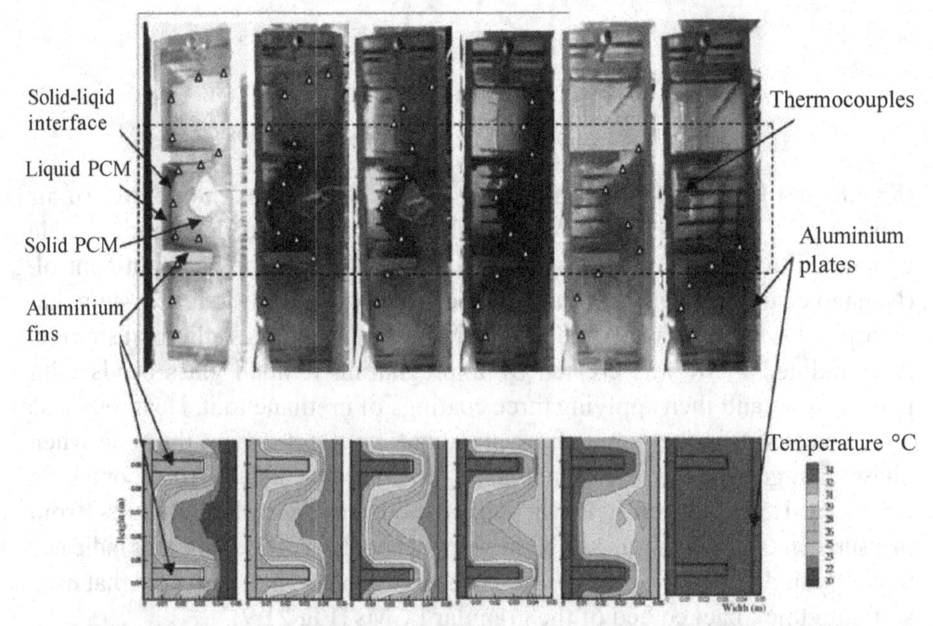

Fig. 21.10: The photographic image of the studied system(Huang et al. 2006)

change. Similar kind of manyresearch study had been also done in the similar research theme. Hasan et al. (2010) also conducted the similar kind of research work and used five PCM for thermal regulation enhancement of building integrated photovoltaic. The study showed that the thermal regulation performance of a PCM depends on the heat capacity of PCM and thermal conductivity of PCM and overall PV/PCM system (Fig. 21.10).

(Aroul and Velraj, 2010) conducted a research work on the thermal performance of a eutectic PCM (48% $CaCl_2$ + 4.3% NaCl +0.4% KCl +47.3% H_2O) for thermal management in a residential building and found it promising (Fig. 21.11).

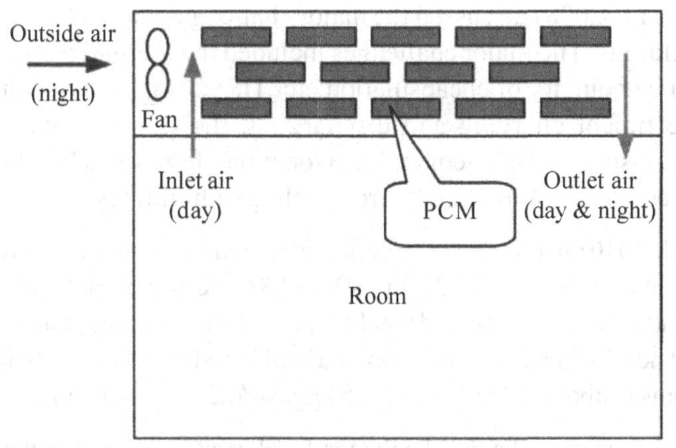

Fig. 21.11: The diagram of the proposed system (Aroul and Velraj 2010)

(Ravikumar and Srinivasan 2003) conducted tests on transient heat transmissions across different roof structures and found that when PCM is installed in withering course region (WC– a mixture of broken bricks and lime-mortar), the nearly uniform roof-bottom surface temperature was maintained. (Aghshenaskashaniand Pasdarshahri 2009) performed two dimensional model for simulation and analysis of PCMs in order to minimize energy consumption in buildings. The effect of using PCM in integrated brick was investigated numerically. The experiment showed that by utilizing PCM in brick the maximum inlet flux may reduce to about 32.8 % depending upon PCM quantity. It also showed that to obtain maximum performance the PCM should be located near outdoor. (Zhou et al. 2009) numerically investigated the effect of shape stabilized PCMs (SSPCM) plates combined with night ventilation in summer. These plates are applied as inner linings of walls and the ceiling of a building without air-conditioning (Fig. 21.12). Unsteady simulation is performed using a verified enthalpy model. The results show that SSPCM plates could decrease the daily maximum temperature by up to 2°C due to coolness, which stored at night.

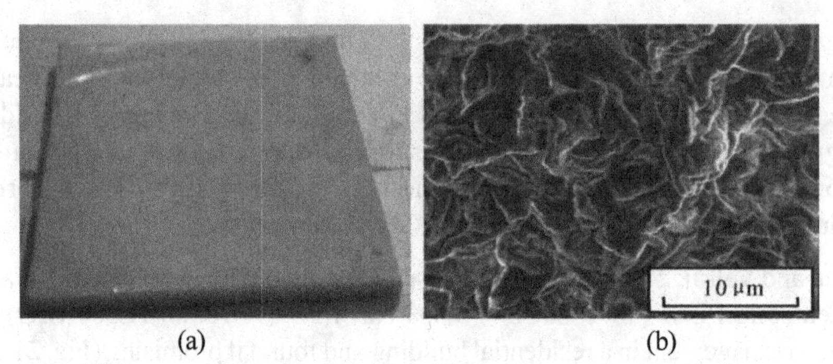

(a) (b)

Fig. 21.12: (a) PCM plate (b) SEM image (Zhou et al. 2009)

(Pasupathy et al., 2008) discussed the major challenges and fact on the concept of green buildings. The major challenges included the thermal resistance of air and PCM, the geometry of encapsulation etc. They also discussed the method of energy efficient charging and discharging, the effect of phase change temperature, insulation and geographical location of PCMs in buildings. They also discussed the PCM used in the free cooling of buildings.

(Castell et al. 2010) had done an experimental study of using PCM for passive cooling. Commercial PCMs RT-27 and SP-25 A8 (macro encapsulated) were used during the study. The result showed that PCM could minimize the peak temperature upto 1 °C. The electrical consumption could also reduce up to 15% in summer which amounts to about reduction of 1-1.5 kg/year/m2 of carbon dioxide emission.

(Diaconu 2011) proposed a PCM enhanced wall system in their paper in which the effect of occupancy pattern and ventilation on the energy savings potential of the wall system is studied. The result showed that the occupancy pattern influence the value of PCM melting point and ventilation and its pattern reduced the relative pattern of energy saving.(Karlessi et al. 2011) investigated the performance of organic PCMs when incorporated in coatings for buildings and urban fabric. PCM coatings can be up to 8°C cooler than common coatings of the same color presenting a surface temperature reduction of 12% (Fig. 21.13).

Fig. 21.13: PCM coatings (Karlessi et al. 2011)

(Stritih 2004) developed a cheap thermal storage system using paraffin wax as a thermal energy storage medium with an ordinary air/water heat exchanger to store and utilize the heat or cold. His experiment with Terhell paraffin showed good heating and cooling characteristics.

(Xu et al. 2013) studied the thermal performance of a room using a floor with shape stabilized PCM. The PCM used was 70 wt% paraffin as dispersed PCM, 15 wt% polyethylene and 15 wt% of styrene-butadiene-styrene block supporting material copolymer. The experiments showed that the mean indoor temperature of a room with the PCM floor was about 2°C higher than that of the room without PCM floor and the indoor temperature swing range was narrowed greatly. This manifests that applying shape stabilized PCM in the room suitably can increase the thermal comfort degree and save space heating energy in winter.

Solar Cooker

(Sharma et al. 2000) designed and developed a cylindrical PCM storage unit for a solar cooker to store solar energy during sunshine hours. Cooking experiments were conducted with different loads and loading times during summer and winter seasons. The experimental results showed that if a PCM having melting temperature between 105°C and 110°C is used; the cooking with the present design will be possible even during the night (Fig. 21.14).

(Sharma and Sagara 2004) designed and developed PCM storage unit to store solar energy during fine sunshine hours. Commercial grade Erythritol (melting point 118°C, latent heat of fusion 339.8 kJ/kg) was used as a latent heat storage material and it was observed that noon cooking did not affect the evening cooking, and evening cooking using PCM heat storage was found faster than noon cooking. In summers, PCM temperatures were found more than 110°C at the time of evening cooking which is higher than the boiling point of water and sufficient for cooking most kind of foods. Hence, Erythritol may be used as a promising PCM for solar cooking.

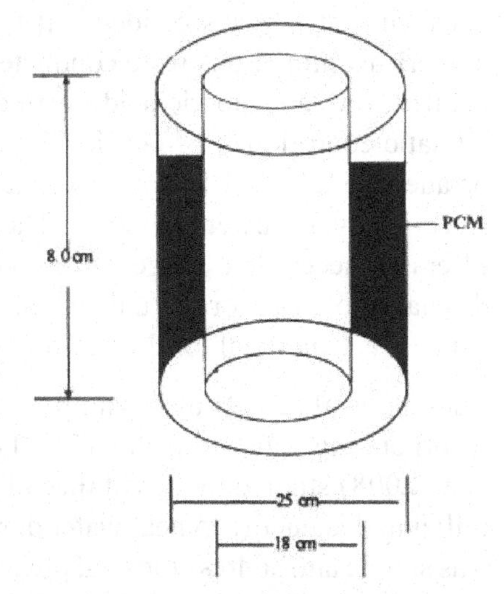

Fig. 21.14: The diagram of the latent heat storage unit(Sharma et al. 2000)

(Escudero et al. 2010) carried out research and development of various types of solar cookers, in particular, the storage type concentrating solar cookers using PCM A-164 and observed that Box type direct solar cookers and Solar steam cooking using parabolic concentrators are also successfully being utilized for community-levellarge-scale noon meal cooking. (El-Sebaii et al. 2011) investigated the influence of the melting/solidification fast thermal cycling of commercial grade $MgCl_2.6H_2O$ on its thermo-physical properties; such as melting point and latent heat of fusion, which is to be used as a storage medium inside solar cookers. They showed that it was a promising material for the solar cooker with the further advantage that it suffered little supercooling.

Solar water heating systems

(Long and Zhu 2008) developed an air source heat pump water heater with PCM for thermal storage to take advantage of off-peak electrical energy. Their work provides guidelines for air source heat pump water heater with PCM for thermal storage.(Talmatsky and Kribus 2008) experimentedon the heat storage tank with and without PCM. Annual simulations were carried out for different sites, load profiles, different PCM volume fractions, and different kinds of PCM. The results of all simulation scenarios indicate that, contrary to expectations, the use of PCM in the storage tank does not yield a significant benefit in energy provided to the end-user. The main reason for this undesirable effect is found to be increased heat losses during nighttime due to reheating of the water by the PCM.

(Qarnia 2009) developed a theoretical model based on the energy equation to predict the thermal behavior and performance of a solar latent heat storage unit consisting of a series of identical tubes embedded in the PCM. A series of numerical simulations were conducted for three kinds of PCM (n-octadecane, paraffin wax and stearic acid) to find the optimum design under the summer climatic conditions of Marrakech city. This study showed that the use of n-octadecane as PCM is not beneficial because the outlet temperature of hot water is never greater than 28°C. The results also showed that the Stearic acid offers an acceptable range of the outlet temperature of hot water and fully discharge of the storage unit for an optimum mass flow rate of water, and hence it is beneficial for the heating water application.

(Shukla et al. 2009) also summarized the thermal energy storage methods incorporating with and without PCM for use in solar water heaters. (Padmaraju et al. 2008) studied the feasibility of storing solar energy using the PCM and utilizing this energy to heat water during night time. Their experimental setup was simple and address the dual objective of providing the heated water during the day as well as night (Fig. 21.15). It consisted of two setups one equipped with the solar water heating system installed to provide water during the daytime and the other unit consists of a heat storage unit with PCM. The water heating system provides the hot water during the daytime and the storage unit absorbs heat during the daytime and the absorbed heat was then utilized to heat water during the night time and overcast condition. For the PCM they used paraffin wax filled in the aluminum containers.

Fig. 21.15: The photograph of the experimental setup (Padmaraju et al. 2008)

(Reddy et al. 2012) studied the different PCM viz. Paraffin and Stearic acid by varying the amount the heat transfer fluid (THF) rates and the size of the spherical capsules (68mm, 58mm, 38mm in diameter). The experiment was performed both during charging and discharging process. The result showed that the performance of the 38mm capsule size was better than the other capsules. The experimental setup has also shown in Fig. 21.16.

Fig. 21.16: The photograph of the experimental setup (Reddy et al. 2012)

Solar air heating

Takeda et al. (2003) constructed a computer simulation program for the PCM floor supply air conditioning system. Suitable phase change temperature T_m for cooling is chosen among six conditions when thermal storage time is 8 hours. As the result, the air conditioner load in the daytime can be minimized when T_m is between 19°C and 22°C. Thermal storage amount in the night time is increased by 27.9% by utilizing the PCM packed bed when thermal storage time is 10 hour. (Enibe 2002) presented design, construction and performance evaluation of a passive solar powered air heating system consists of a single-glazed flat plate solar collector integrated with a PCM heat storage system (Fig. 21.17). The PCM was prepared in modules, with the modules equi-spaced across the absorber plate. The spaces between the module pairs serve as the air heating channels, the channels being connected to the common air inlet and discharge headers at the ambient temperature range of 19–41°C, and a daily global irradiation range of 4.9–19.9 MJ m^2. Peak temperature rise of the heated air was about 15 K, while the maximum airflow rate and peak cumulative useful efficiency were about 0.058 kg s^{-1} and 22%, which respectively show that the system can be operated successfully for crop drying applications.

Fig. 21.17: A. Energy storage and heating system B. Heated space (Enibe, 2002)

(Diaconu et al. 2010) experimentally investigated heat storage properties and heat transfer characteristics in order to assess its suitability for the integration into a low-temperature heat storage system for solar air conditioning applications. An experimental set-up was built in order to quantify the natural convection heat transfer occurring from a vertical helically coiled tube immersed in the phase change material slurry (Fig. 21.18). It was found that the values of the heat transfer coefficient for the PCM slurry were higher than for water, under identical temperature conditions inside the phase change interval.

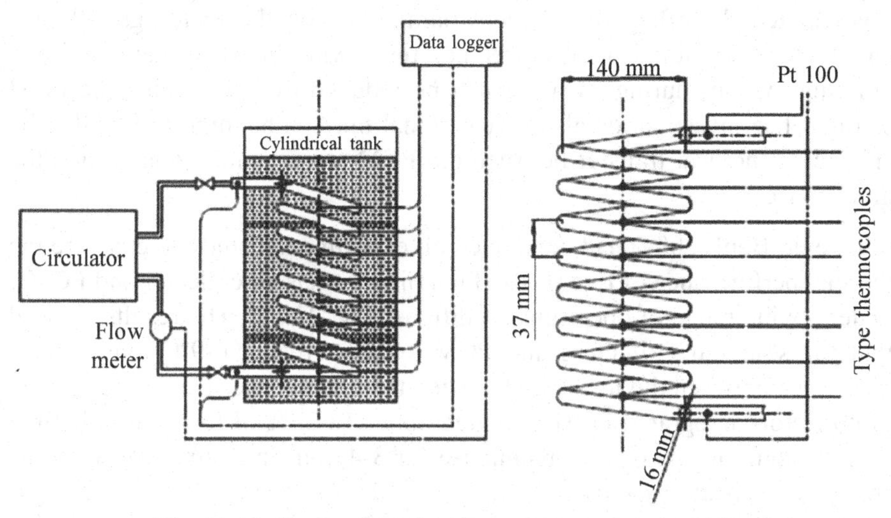

Fig. 21.18: The Experimental Setup (Diaconu et al. 2010)

(Hed and Bellander, 2006) developed the mathematical model for the PCM based heat exchanger. The model was then validated with the prototype of the heat exchanger. The result was very promising and agree with the mathematical model. (Salwa et al. 2013) conducted the thermal performance of a solar air heater collector using packed bed spherical capsules with PCM (Fig. 21.19). In the current experiment, both the energetics and exegetic efficiencies were studied. The daily efficiency energy efficiency was reported between 32% to 45% while the daily exergy efficiency was between 13% to 25%.

Fig. 21.19: The schematic diagram of the experimental setup (Salwa et al. 2013)

Solar greenhouse

Boulard et al. 1990 tested a greenhouse with a PCM heat storage system containing a quasi-eutectic mixture with a classical lettuce-tomato rotation. Experiments showed that in the south of France, this heat storage system can keep a greenhouse temperature roughly 10°C higher than outside during typical nights of March and April. (Najjar and Hasan 2008) developed a mathematical model for the storage material for greenhouse application. They solved the coupled models using numerical methods and java code programming. They investigated the effect of different parameters on the inside greenhouse temperature and found that the temperature swing between maximum and minimum values during 24 hours can be reduced by 3-5°C using the PCM storage. The investigation also revealed that this can be improved further by enhancing the heat transfer between the PCM storage and the air inside the greenhouse.

(Hu¨seyin Benli 2009) had conducted an experimental study to evaluate the thermal performance of five types of ten pieces solar air collector and PCMs, under a wide range of operating conditions (Fig. 21.20). The results showed that the solar air collectors and PCM system created 6-9°C temperature difference between the inside and outside the greenhouse. The proposed size of collectors integrated PCM provided about 18-23% of total daily thermal energy requirements of the greenhouse for 3-4 hour in comparison with the conventional heating device.

(a)

(b)

Fig. 21.20: The Experimental Setup(Hu¨seyin Benli, 2009)

(Boulard et al. 1990) developed a quasi-eutectic mixture used as a PCM in a greenhouse. Authors concluded that such a heat storage system, when used in the South of France, can keep a greenhouse roughly 10°C higher than outside during typical nights of March or April.

Conclusion

As the energy resources are depleting and energy prices are going up our look for the alternative energy resources can be stopped at the solar energy with the latent heat storage concept through PCM. Many latent heat storage devices with their technological advancement have been discussed. This chapter also presents the state of art of various possible PCM-based thermal energy storage systems with different applications. Those technologies are very beneficial for the energy conservation. The heat storage applications are used as a part of building applications, solar cookers, solar water heating systems, solar air heating systems, solar greenhouse etc. It is obvious that these research results will help to diversify applications of PCMs in general.

Acknowledgment

Authors are highly thankful to Council of Science and Technology, UP (Reference No. CST 3012-dt.26-12-2016) for providing the research grants to carry out the work at the institute.

References

Aroul, V.A., Velraj, R., 2010. Review on free cooling of buildings using phase change materials. Renewable and Sustainable Energy Reviews 14, 2819–2829. https://doi.org/10.1016/j.rser.2010.07.004

Castell, A., Martorell, I., Medrano, M., Pérez, G., Cabeza, L.F., 2010. Experimental study of using PCM in brick constructive solutions for passive cooling. Energy and Buildings 42, 534–540. https://doi.org/10.1016/j.enbuild.2009.10.022

Diaconu, B.M., 2011. Thermal energy savings in buildings with PCM-enhanced envelope/ : Influence of occupancy pattern and ventilation. Energy & Buildings 43, 101–107. https://doi.org/10.1016/j.enbuild.2010.08.019

Diaconu, B.M., Varga, S., Oliveira, A.C., 2010. Experimental assessment of heat storage properties and heat transfer characteristics of a phase change material slurry for air conditioning applications. Applied Energy 87, 620–628. https://doi.org/10.1016/j.apenergy.2009.05.002

Dong, Z., Cui, H., Tang, W., Chen, D., Wen, H., 2016. Development of hollow steel ball macro-encapsulated PCM for thermal energy storage concrete. Materials 9. https://doi.org/10.3390/ma9010059

El-Sebaii, A.A., Al-Heniti, S., Al-Agel, F., Al-Ghamdi, A.A., Al-Marzouki, F., 2011. One thousand thermal cycles of magnesium chloride hexahydrate as a promising PCM for indoor solar cooking. Energy Conversion and Management 52, 1771–1777. https://doi.org/10.1016/j.enconman.2010.10.043

Enibe, S.O., 2002. Performance of a natural circulation solar air heating system with phase change material energy storage 27, 69–86.

Escudero, C., Yates, C.A., Buhl, J., Couzin, I.D., Erban, R., Kevrekidis, I.G, Maini, P.K., 2010. Ergodic directional switching in mobile insect groups. Physical Review E 82, 011926. https://doi.org/10.1103/PhysRevE.82.011926

Fisch, M.N., Kühl, L., 2004. Use of Microencapsulated Phase Change Materials in Office Blocks.

Hasan, A, McCormack, S.J. Humg, M.J. and Norton, B. 2010. Evaluation of phase change material for thermal regulation enhancement of building integrated photovoltaic. Solar Energy 84(9): 1601-1612.

Hed, G., Bellander, R., 2006. Mathematical modelling of PCM air heat exchanger 38, 82–89. https://doi.org/10.1016/j.enbuild.2005.04.002

Hu¨seyin Benli, A.D., 2009. Performance analysis of a latent heat storage system with phase change material for new designed solar collectors in greenhouse heating 83, 2109–2119. https://doi.org/10.1016/j.solener. 2009.07.005

Huang, M.J., Eames, P.C., Norton, B., 2006. Phase change materials for limiting temperature rise in building integrated photovoltaics. Solar Energy 80, 1121–1130. https://doi.org/10.1016/j.solener.2005.10.006

Karlessi, T., Santamouris, M., Synnefa, A., Assimakopoulos, D., Didaskalopoulos, P., Apostolakis, K., 2011. Development and testing of PCM doped cool colored coatings to mitigate urban heat island and cool buildings. Building and Environment 46, 570–576. https://doi.org/10.1016/j.buildenv.2010.09.003

Kondo, T., Ibamoto, T., 2002. Research on Using the PCM for Ceiling Board. presented at IEA ECESIA 2352.

Long, J., Zhu, D., 2008. Numerical and experimental study on heat pump water heater with PCM for thermal storage 40, 666–672. https://doi.org/10.1016/j.enbuild.2007.05.001

Nagano, K., Takeda, S., Mochida, T., Shimakura, K., Nakamura, T., 2006. Study of a floor supply air conditioning system using granular phase change material to augment building mass thermal storage — Heat response in small scale experiments 38, 436–446. https://doi.org/10.1016/j.enbuild.2005.07.010

Najjar, A., Hasan, A., 2008. Modeling of greenhouse with PCM energy storage 49, 3338–3342. https://doi.org/10.1016/j.enconman.2008.04.015

Padmaraju, S.A.V., Viginesh, M., Nallusamy, N., 2008. Comparative study of sensible and latent heat storage systems integrated with solar water heating unit. Renewable Energy & Power Quality Journal 1–6. https://doi.org/10.24084/repqj06.218

Pasupathy, A., Athanasius, L., Velraj, R., 2008. Experimental investigation and numerical simulation analysis on the thermal performance of a building roof incorporating phase change material (PCM) for thermal management 28, 556–565. https://doi.org/10.1016/j.applthermaleng.2007.04.016

Qarnia, H. El, 2009. Numerical analysis of a coupled solar collector latent heat storage unit using various phase change materials for heating the water. Energy Conversion and Management 50, 247–254. https://doi.org/10.1016/j.enconman.2008.09.038

Reddy, R.M., Nallusamy, N., Reddy, K.H., 2012. Experimental studies on phase change material-based thermal energy storage system for solar water heating applications. Journal of Fundamentals of Renewable Energy and Applications 2, 1–6. https://doi.org/10.4303/jfrea/R120314

Rudd, A., 2008. Phase-Change Material Wallboard for Distributed Thermal Storage in Buildings PHASE = CHANGE MATERIAL WALLBOARD iN BUILDINGS.

Sharma, A., Tyagi, V.V., Chen, C.R., Buddhi, D., 2009. Review on thermal energy storage with phase change materials and applications. Renewable and Sustainable Energy Reviews 13, 318–345. https://doi.org/10.1016/j.rser.2007.10.005

Sharma, S.D., Buddhi, D., Sawhney, R.L., Sharma, A., 2000. Design, development and performance evaluation of a latent heat storage unit for evening cooking in a solar cooker. Energy Conversion and Management 41, 1497–1508. https://doi.org/10.1016/S0196-8904(99)00193-4

Sharma, S.D., Kotani, H., Kaneko, Y., Yamanaka, T., Sagara, K., 2007. Design, development of a solar chimney with built-in latent heat storage material for natural ventilation. International Journal of Green Energy 4, 313–324. https://doi.org/10.1080/15435070701332120

Shukla, A., Buddhi, D., Sawhney, R.L., 2009. Solar water heaters with phase change material thermal energy storage medium/ : A review 13, 2119–2125. https://doi.org/10.1016/j.rser.2009.01.024

Stritih, U., 2004. An Experimental Model of Thermal Storage System for Active Heating or Cooling Of Buildings Uroš Stritih University of Ljubljana , Faculty of Mechanical Engineering , PHASE-CHANGE MATERIALS. Annex 17 1–8.

T. Boulard, E.Razafinjohany, A.Baille, A.Jaffrin, B.F., 1990. Performance of a greenhouse heating system with a phase change material The solar greenhouse. Agricultural and Forest Meteorology 52, 303–318.

TeKeda, S., Nagano, K., Mochida, T. and Nakamura, T. (2009) Development of floor supply air conditioning system using granulated phase change materials. Proceedings of Futurestock 2003 9th International Conference on Thermal Energy Storage, Warsaw, pp 317-322.

Talmatsky, E., Kribus, A., 2008. PCM storage for solar DHW: An unfulfilled promise? Solar Energy 82, 861–869. https://doi.org/10.1016/j.solener.2008.04.003

Xu, X., Zhang, Y., Di, H., Lin, K., 2013. Experimental Study on the Thermal Performance of Phase Change Material Floor Combined with Solar Energy. Journal of Solar Energy Engineering 128, 1–12. https://doi.org/10.1115/1.2189866

Zhou, G., Yang, Y., Wang, X., Zhou, S., 2009. Numerical analysis of effect of shape-stabilized phase change material plates in a building combined with night ventilation. Applied Energy 86, 52–59. https://doi.org/10.1016/j.apenergy.2008.03.020

22

Passive and Hybrid Cooling Systems for Building in Hot and Dry Climatic Condition

S.P. Singh and Digvijay Singh

School of Energy and Environmental Studies
Devi Ahliya Viswavidyalaya (DAVV), Indore, Madhya Pradesh, India

Introduction

The thermal performance of a building refers to the process of modeling the energy transfer between a building and its surroundings. For a conditioned building, it estimates the heating and cooling load and hence, the sizing and selection of HVAC equipment can be correctly made. For a non-conditioned building, it calculates temperature variation inside the building over a specified time and helps one to estimate the duration of uncomfortable periods. These quantifications enable one to determine the effectiveness of the design of a building and help in evolving improved designs for realizing energy efficient buildings with comfortable indoor conditions. The lack of proper quantification is one of the reasons why passive solar architecture is not popular among architects. Clients would like to know how much energy might be saved, or the temperature reduced to justify any additional expense or design change. Architects too need to know the relative performance of buildings to choose a suitable alternative. Thus, knowledge of the methods of estimating the performance of buildings is essential to the design of passive solar buildings.

The demand of energy in developing countries is increasing with the increase in the standards of living and population. In India building sector is the second largest consumer of electricity. The electricity demand from household buildings of India will raise 5 fold to 842 TWh by 2030 from the base level of 175 TWh in 2012 having electricity and liquids as the main energy sources, which contributes depletion of fossil fuels, to atmospheric pollution and to climatological changes

(Singh and Bhat, 2018). Passive cooling methods can play a major role in reducing the building demand and occupants dissatisfaction with indoor comfort conditions, it involves the natural process for achieving the thermal comfort in the building without the use of electricity or very low electricity through radiation, conduction and convection, the natural environmental heat sinks used in the process are given in Table 22.1, especially for the climate of hot-dry zones characterized by high temperatures (40-50°C in summer), with sharp variations in both diurnal and seasonal temperatures energy required for cooling is higher than the heating. Based on various climatic data the country has been divided into five climatic zone (Table 22.2) and a number of passive cooling technique have been provided in Fig. 22.1 (Bansal et al. 1988).

Table 22.1: Classification of heating and cooling concepts.

Technique	Heat Transfer Mode	Sink
Radiative cooling	Radiation	Sky
Evaporative cooling	Evaporation	Air
Convective cooling	Convection	Air
Ground cooling	Conduction	Earth

Table 22.2: Classification of climatic zones

Climate	Mean monthly temperature	Relative humidity (%)	Precipitation (months)	Number of clear days	Example
Hot and dry	> 30	<55	<5	>20	Jodhpur
Warm and humid	>30	>55	>5	<20	Bombay
Moderate	25-30	<75	<5	<20	Bangalore
Cold and cloudy	<25	<55	<5	<20	Srinagar
Composite	This applies when ≥6 months do not fall within any of the above categories				New Delhi

Passive cooling strategies

To avoid overheating and create thermal comfort conditions in the interior of buildings during summer, the cooling strategies should be designed at these levels (Asimakopoulos 2013; Givoni 1994 and NITI Aayog 2015)

- Prevention of heat gains in the building
- Reduce Solar Gain through Walls and roof
- Minimize Conduction Gain
- Maximize Air Movement on external Surfaces (walls and roof) to increase convective losses
- Maximize Selective Ventilation
- Minimize Infiltration to reduce cooling Losses

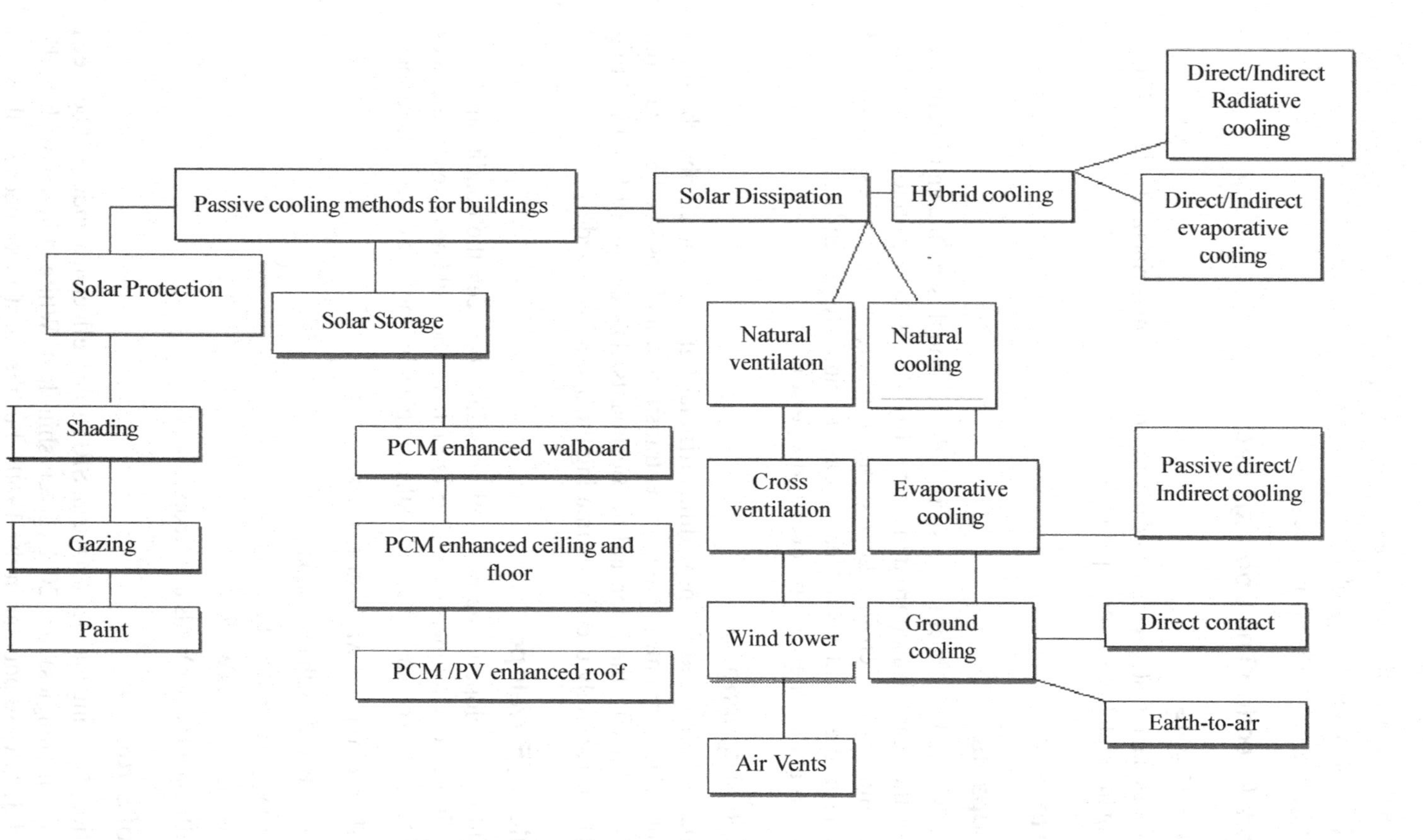

Fig. 22.1: Classification of Passive cooling techniques

- Optimize building Orientation
- Choose Appropriate Colors
- Shading of building components

Factors Affecting The Cooling Load

From outdoors

Heat from solar radiation conducted through building materials and solar radiation entering through transparent openings.

From indoor

• Occupants

Human bodies loses heat to the surrounding through sensible and latent heat process, the heat released from the surface of skin with respect to surrounding temperature called sensible heat while heat released from respiration and sweating is the latent heat. Heat released primarily depends on the indoor conditions and on the activity level of the person.

• Artificial lighting

Luminaries were used for providing artificial lighting, it itself absorbs some amount of heat, remaining amount is transmitted and absorbed by the room surfaces primarily, the floor and the occupants. The use of artificial lighting needs to be carefully used by estimating the operating schedule.

• Building Envelope

A building envelope is the physical boundary between the conditioned and unconditioned environment enclosing a structure. The building envelope includes the materials that comprise mainly wall, roof, glazing, door and other penetrations.

The different type of walls structures are-

i. Cavity wall with insulation
ii. Cavity wall without insulation
iii. Hollow Blocks Walls
iv. Radiant Barrier Walls and Roofs

• Roof Structures

Traditional roofing materials have an SRI (Solar reflectance index) of between 5% (brown shingles) and 20% (green shingles). White shingles with SRI's around 35% were popular in the 1960s, but they lost favor because they get dirty easily. There must be optimum roof structures with minimum discomfort index for different climatic zones.

Solar shading

It's a technique used for controlling solar radiation for optimizing heat gain and daylight requirement. A window is important part of building for accessing all heat gain, air, and visible light. Window must always be shaded from the direct solar component and often so from the diffuse and reflected components. The different criteria of shading of buildings for various climatic zones have been given below.

Shading by overhangs, louvers and awnings

Shading with the help of retrofitting to block unwanted heat gain, which reduces the cooling load of air-conditioning system,

- Overhangs of chajjas give protection to the wall and prevention against sun and rain.
- Louvers are adjustable or can be fixed type, obstruct air movement and provide shading to the building, optical shutters consists of three layers of transparent sheets and one layer of cloud gel. It is opaque at high temperatures used for reducing cooling load and preventing overheating in green houses and solar-collector systems.
- Awnings are attached to external wall of a building, it's a sheet of canvas or some other material stretched on a frame.

Shading by trees and vegetation

Shading by trees and vegetation are economical source for reducing energy loads of the buildings and also the environment load, they can be used for shading of roof, walls and windows, chosen on the basis of their growth environment to provide desired degree of shading. It was found that large trees can provide up to 70% of shade during spring and autumn season in comparison to younger ones (Gomez-Munoz et al. 2010).

In case of roof a cover of deciduous plant and creepers can be used, the evaporation from the leaf surface brings down the temperature of roof. Shading and insulation for walls can be provided by plants that stay to the wall, such as English ivy, or by plants supported by the wall, such as jasmine.

Shading by textured surfaces

Highly textured walls and closely packed inverted earthen pots on the roof increase the surface area with the increased outer surface coefficient, which permits the sunlit surface to stay cooler as well as to cool down faster at night. Such a heavily insulated walls and roofs need less shading.

Thermal storage cooling

To reduce cooling load of the buildings, Phase Change Material (PCM) can be used as a Thermal Energy Storage (TES). PCM allows large amounts of energy to be stored in relatively small volume. It mimics effect of thermal mass. It absorbs heat in endothermic process and changes phase from solid to liquid and releases heat in an exothermic process.

PCM/PV enhanced roof

In order to ease thermal bridging and reduce roof-generated thermal loads, Kosney et al. (2012) developed new roofing technology utilizing amorphous silicon PV laminates integrated with the metal roof panels. During summer, the PCM was expected to act as a heat sink, reducing the heat gained by the attic and consequently, lowering the building cooling-loads. The thermal performance of the experimental roof was compared to a control attic with a conventional asphalt shingle roof (Souayfane et al. 2016). The PV-PCM attic had a 30% reduction in roof-generated heating loads compared to a conventional shingle attic during the cooling season, the attic generated cooling loads from the PV-PCM attic were about 55% lower than that of the shingle attic. In addition, about 90% reductions in peak daytime roof heat fluxes were observed with the PV-PCM roof (Fig. 22.2).

Fig. 22.2: Shingle roof and photo-voltaic (PV)-phase change material (PCM) roof – on the left; open air cavity in the PV-PCM roof – on the right side.

PCM enhanced ceiling

Singh and Bhat (2018) developed a false ceiling gypsum board prepared by placing phase change material in small cells in the form of poly-bags. These poly-bags sandwiched in two layers of the gypsum. The gypsum board used for present investigation has two layers of phase change material with different melting points. This gypsum board is intended to be used as false ceiling and hence can easily be used with existing building, it shows the maximum reduction in cooling load for a conditioned building (Fig. 22.3).

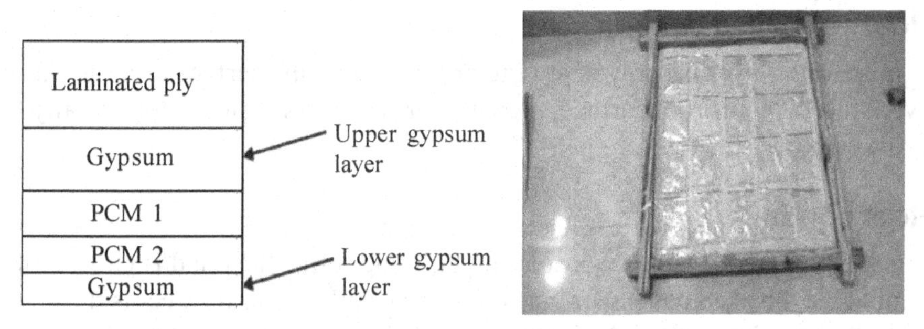

Fig. 22.3: (a) Structure of test cell (b) Photograph of dual PCM gypsum board in making

Evaporative cooling

Evaporative cooling is a process that uses the effect of evaporation as a natural heat sink. Sensible heat from the air is absorbed to be used as latent heat necessary to evaporate water. The amount of sensible heat absorbed depends on the amount of water that can be evaporated (Fig. 22.4).

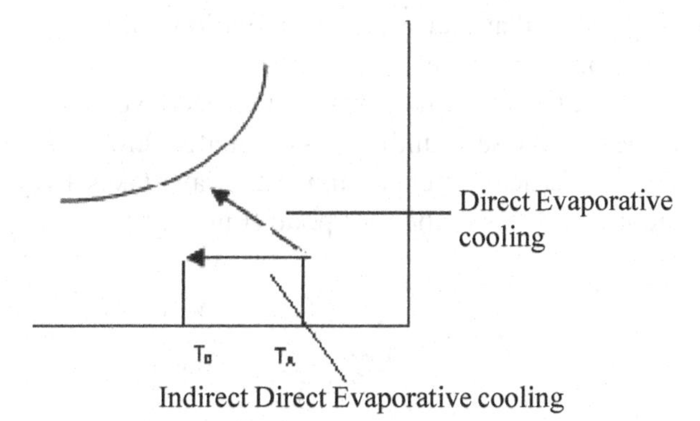

Fig. 22.4: Process of psychometric chart

Passive direct/Indirect cooling systems

Passive direct cool systems

Air humidification and cooling by evapotranspiration of plants and the use of free water surfaces, like pools and streams, is a passive direct technique.

Passive indirect cooling systems

It uses natural phenomena, having environment as source and sink, however, it may contain small power fan and pumps, reducing the air temperature of the building up to 9ÚC. These systems consist of:

- Roof sprinkling
- Roof ponds
- Moving water films

Roof sprinkling

A mist water layer is sprayed at regular interval on the surface roof, as water evaporates it cools roof surface, these system can be operated automatically to avoid run off.

Roof ponds

It gets integrated to the non-insulating building roofs, the pond is covered by removable, or fixed cover. During summer the ambient air flow cools the pond as well as roof structure which act as a heat sink to the building, the necessary condition for applying this technique efficiently, is that the wet-bulb temperature of the air should be lower than 20°C.

Cool-Pool system

A cool-pool system for passive cooling of a non-conditioned building is given by Sodha ct al. (1985). The system consists of a water pond over the roof which is shaded in such a way that incident solar radiation does not reach the roof surface and the pond loses heat by convection and evaporation to its surroundings. The pool water, which gets cooled by convective, Radiative and evaporative heat loss to the surroundings, is circulated through the interior of a building by means of vertical water column. This water takes away some of the interior heat and then returns to the roof pond (Fig. 22.5).

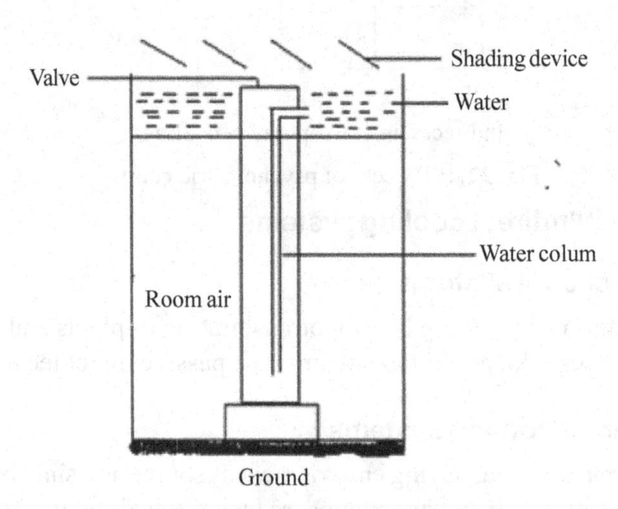

Fig. 22.5: Cool-pool system

Ventilation techniques

It is the initiation of ambient air into a space, either natural or mechanically depending upon the outdoor condition to improve the thermal comfort and air quality of the space. In case of natural ventilation, natural process occurs due to effect of wind and buoyancy driven force where the air flowing next to a surface carries away heat, provided it is at a lower temperature than the surface, resulting from various openings of the building (doors, window, etc.) also from the cracked structure of the building. Ventilation from cracked structure is called air-infiltration, so the building must be air tight to avoid infiltration.

Stack Effect

It can exhaust air from a building by the action of natural convection as shown in Fig. 22.6

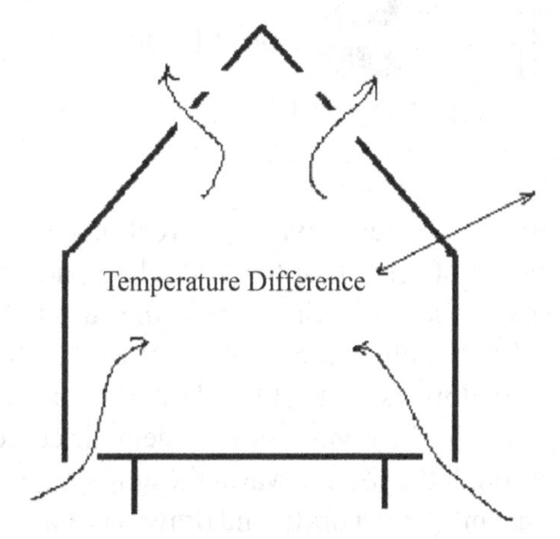

Fig. 22.6: Stack effect will exhaust air outside when temperature of indoor is higher than outdoor

Solar chimney

It's a tall black painted cavity passive ventilation system. The air inside the cavity gets heated from the solar radiation. Due to the buoyancy force its gets rises up out of the chimney allowing the fresh air from outside of the building at a quick rate. Solar chimneys when combined with evaporative cooling device improve the comfort condition during daytime (Fig. 22.7).

It can be classified on the basis of-

- Vertical solar chimney
- Inclined Solar chimney
- Trombe wall solar chimney
- Roof Solar Chimney
- Single glazing
- Double glazing

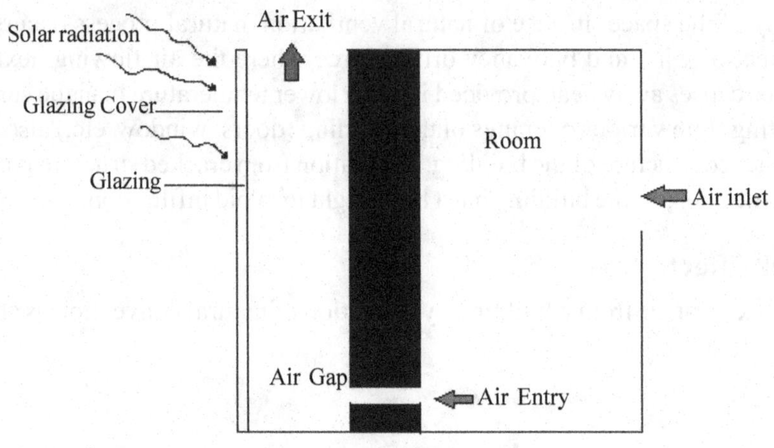

Fig. 22.7: Solar Chimney

Cool tower

These are the passive evaporative coolers having different shapes and structures. It's a reverse of solar chimney, at the top of tower cooling pads are provided, water is sprayed on these pads, the warm air enters in the pads become denser, and flows down filling building with cool air naturally, although fans may be used to reduce the size of the towers. After the whole day process the tower becomes warm in the evening. In the night, cool ambient air comes in contact with bottom of the tower through the rooms, warm the ambient air in the tower. Warm air moves up, creating an upward draft, and draws cool night air through the doors and windows into the building (Fig. 22.8 & 22.9).

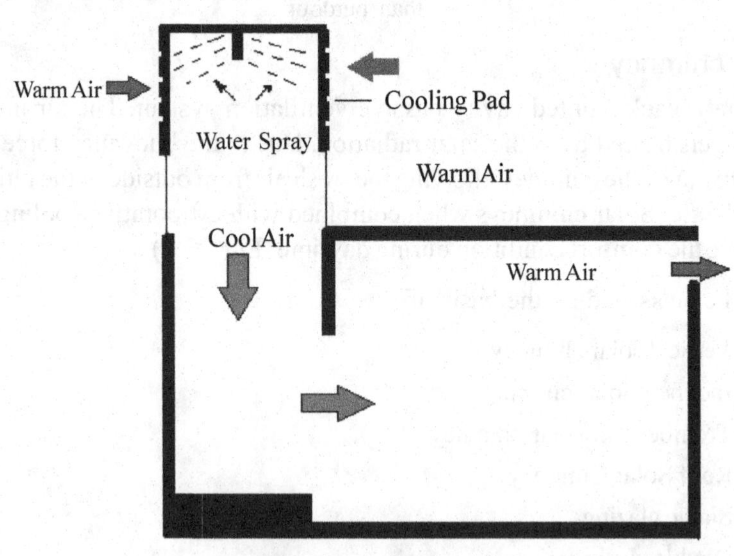

Fig. 22.8: Cool tower

Fig. 22.9: Cool tower architecture at School of Energy and Environmental Studies, Indore

Air vents

Air vents are used for cooling air in hot and dry climate, replaces the cool tower where the dusty wind makes cool tower not viable for such climate. It's a domed/cylindrical shape roof structure with the cap, the system works on the principle of cooling by induced ventilation.

Earth cooling

Principle of earth cooling is temperature remains equal to annual average ambient temperature or annual average Sol-air temperature (Tsol-air) ~ 22 °C to 24 °C.

Direct contact

In direct contact method also known as earth sheltering, the conductive heat exchange process is increased by making direct contact of building envelope with ground as shown in Fig. 10, one side is completely buried in the ground. This method helps in achieving thermal inertia, decreasing temperature swings inside the building.

Earth-to-air

The principle of a modem system is shown in Fig. 23.10, ambient air is sucked by means of a fan from the ambient and enters the building through the buried pipe. During summer, as has previously been explained, since the ground temperature is lower than the ambient air, the air temperature at the outlet of the exchanger is lower than the temperature at the inlet. The opposite phenomenon occurs during winter. Earth-to-air heat exchangers can be applied in either an open-loop or a closed loop circuit. In a closed-loop circuit, both inlet and outlet are located inside the building. Plastic, concrete or metallic pipes are used in modem applications. They are used to cool the ambient air before injecting it into the building, or the indoor air if the system is used in a closed loop. The temperature decrease of the air depends upon the inlet air temperature, the ground temperature at the depth of the exchanger, the thermal conductivity of the pipes and the thermal diffusivity of the soil, as well as the air velocity and pipe dimensions. Detailed calculations are needed to optimize such a system (Sodha et al. 1985).

A large earth-air tunnel system is developed and evaluated at one of the hospitals in India as shown in the Fig. 22.10. A simple theoretical model is developed to validate the experimental measurements. An 80m length of the tunnel with a cross-sectional area 0.528 m^2 and with an air velocity of 4.89 m s^{-1} is found to have a cooling capacity of approx 512 kWh.

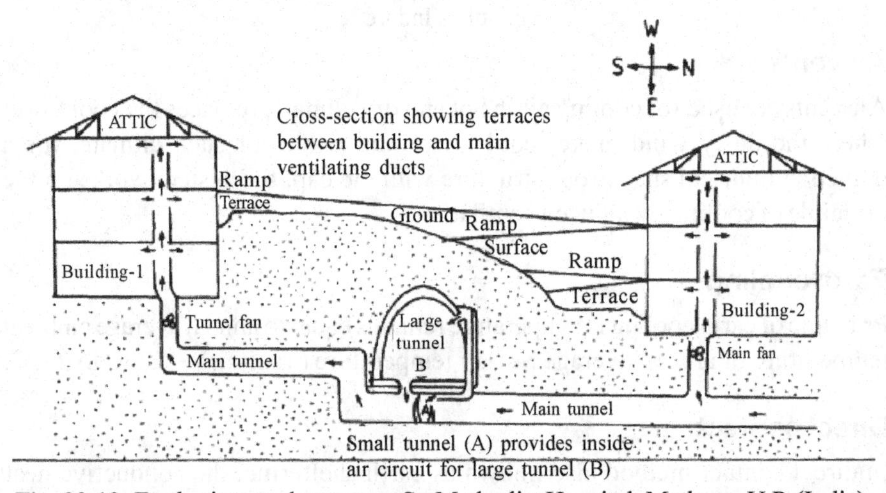

Fig. 22.10: Earth-air tunnel system at St. Methodist Hospital, Mathura, U.P. (India).

Rules of thumb for ground-cooling applications Earth-to-air heat exchanger systems should not be installed prior to specific calculations, at least with the simplified method presented here. However some rules of thumb can be given:

- The length of the exchanger should be at least 10 m
- The diameter of the exchanger should range between 0.2 and 0.3 m.
- The depth of the exchanger should range between 1.5 and 3 m.
- The air velocity through the buried pipe should range between 4 and 8 m.

Radiative cooling

It's based on the heat loss by long-wave radiation emission from a body towards another body of lower temperature, which plays the role of the heat sink. In the case of buildings the cooled body is the building and the heat sink is the sky, since the sky temperature is lower than the temperatures of most of the objects upon earth

Movable insulation

It's the insulation material that can be moved easily over the roof of the building, it exposes the roof mass to the sky during night and insulates it during the daytime from the hotter ambient environment. In winter, these insulation layers one exposed to sun and insulated at night to decreased Radiative heat losses. The major disadvantage of such systems is that there is no liquid fluid is present, for multistory building is only used for space directly upon the roof only (Fig. 22.11).

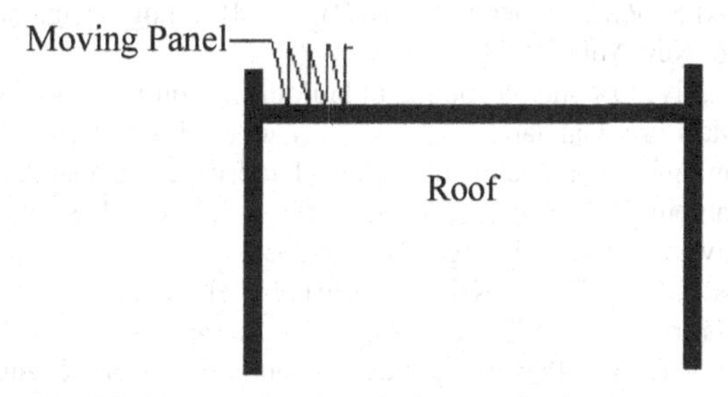

Fig. 22.11: Moving Insulation Radiative system

Movable thermal mass

It consists of a movable insulation device below which there is water pond, a gap present between pond and roof for canalization of water. This system is operated as follows: the pond is filled with water at night and exposed to the sky to be cooled by radiation. In the morning the pond is covered and the cooled water is drained from the pond and circulates between the insulation and the thermal mass of the roof, dissipating heat from the space below (Fig. 22.12).

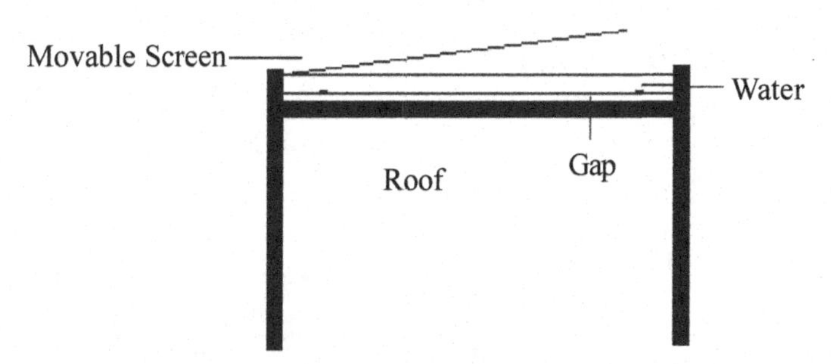

Fig. 22.12: Movable Thermal mass system

Conclusion

- Passive cooling techniques are favorable for hot and dry climates.
- For humid climate only ventilation and indirect cooling will be helpful.

- The first strategy in summer for comfort is to avoid heat.
- Passive cooling can be achieved by a number of ways. The key is to take advantage of the properties of materials and to perhaps go against the gain of traditional construction.
- A cost/benefit analysis may be an important way to demonstrate to a client that such technologies are not only possible but also preferable.

References

Aayog, N.I.T.I. (2015). A Report on energy efficiency and energy mix in the Indian energy system (2030), Using India Energy Security Scenarios, 2047.

Asimakopoulos, D. (2013). Passive cooling of buildings. Routledge.

Bansal N. K., Sodha M.S., P.K., Kumar A., "Solar Passive: Building Design Science", Pergamum Press, Oxford, New York, 1988.

Givoni, B. (1994). Passive low energy cooling of buildings. John Wiley & Sons.

Gomez-Muñoz, V. M., Porta-Gándara, M. A. and Fernández, J. L. (2010). Effect of tree shades in urban planning in hot-arid climatic regions. Landscape & Urban Planning 94: 149-157.

Kolokotsa, D., Santamouris, M., Synnefa, A. and Karlessi, T. 2012. "Passive solar Architecture," Comprehensive. Renewable Energy (3): 637-665.

Koœny, J., Biswas, K., Miller, W. and Kriner, S. (2012). Field thermal performance of naturally ventilated solar roof with PCM heat sink. Solar Energy 86(9):2504-2514.

Singh, S.P. and Bhat, V. (2018). Performance evaluation of dual phase change material gypsum board for the reduction of temperature swings in a building prototype in composite climate. Energy and Buildings 159:191-200.

Sodha, M.S., Sharma, A.K., Singh, S.P., Bansal, N.K. and Kumar, A. (1985). Evaluation of an earth—air tunnel system for cooling/heating of a hospital complex. Building and Environment 20(2):115-122.

Sodha, M. S., Singh, S. P. and Kumar, A. (1985). Thermal performance of a cool-pool system for passive cooling of a non-conditioned building. Building and Environment 20(4):233-240.

Souayfane, F., Fardoun, F. and Biwole, P. H. (2016). Phase change materials (PCM) for cooling applications in buildings: A review. Energy and Buildings 129:396-431.

23

Solar Drying in Food Processing and Effect on Quality Parameters

Soma Srivastava and Dilip Jain

ICAR-Central Arid Zone Research Institute, Jodhpur, Rajasthan, India

Introduction

Considering the accountable post-harvest losses which is approximately 30-35 percent, food processing becomes an extremely important sector for preventing the losses and feeding the growing population. The Indian food industry is poised for huge growth, as currently valued at US$ 1.3 billion with compound annual growth rate of 20 per cent per year, increasing its contribution to world food trade every year. In India, the food sector has emerged as a high-growth and high-profit sector due to its immense potential for value addition, particularly within the food processing industry. Increasing population of the country and high cost of fuels have also created a demand for new alternate sources of energies for post-harvest processing of foods. Solar food processing technologies may become a viable option that provides good quality foods at low or no additional fuel costs. The energy policy of the developing countries now focuses on the clean renewable energy to create a further reduction of the petroleum import and to alter the utilization of petroleum energy toward the utilization of solar energy. It is an interesting fact that most of the developing countries fall in the climatic zone where the insolation is considerably higher than the world average of 3.82 kWh m^{-2} (Majumdar 2015). The daily horizontal insolation of India is approximately 5.80 kWh m^{-2} (Visavale 2009, Visavale et al. 2011). Solar dyers are an alternative to traditional drying technique and a contribution towards the solution of the problems of the open sun drying from perspective of renewable and clean energy. The International Energy Agency (IEA) declared in its 2011 solar energy perspectives executive summary (SEP 2011): "Solar energy offers a clean, climate-friendly, very abundant and inexhaustible energy resource to mankind, relatively wellspread over the globe. Its availability is greater in warm and sunny countries that will experience most of the world's population and economic growth over the next decades.

Solar food processing is an emerging technology that provides good quality foods at low or no additional fuel costs. Solar drying of agricultural products in enclosed structures by forced convection is an attractive way of reducing post-harvest losses and low quality of dried products associated with traditional open sundrying methods (Jain and Tiwari 2003). A number of solar dryers, collectors and concentrators are currently being utilized for the purpose of food processing and value addition. Solar cookers, solar ovens and solar cabinet dryer with forced circulation has been used for dehydration and development of value added products from locally grown fruits, vegetables, leafy greens and forest produce. In many rural areas in developing countries, gridconnected electricity and supplies of other non-renewable sources of energy are either unavailable, unreliable or, too expensive (Jain 2005, Jain and Pathare 2007). In such conditions, solar dryers appear increasingly attractive as commercial propositions (Mekhilefa et al. 2011, Xingxing et al. 2012). It can also create employment opportunities among the rural population, especially women.

Availability of solar radiation in India

India is ideally suited for harnessing solar energy to meet its energy needs. The daily average solar energy incidence across the country varies from 4 to 7 kWh m^{-2}, and in general there are 250 to 300 days of sunshine per year in most parts of the country. Solar dryers can be utilized for 250–300 days in a year in most parts of the country. Even though there is a gradient in the radiation received through various parts of the country, India is divided broadly into five zones for practical purposes: Eastern Zone (3.5–4.0 kWh m^{-2}); Himalayan Zone (4.0–4.6 kWh m^{-2}); Northern Zone (4.6–5.2 kWh m^{-2}); Southern-Middle Zone (5.2–5.8 kWh m^{-2}) and Western Zone (5.8–8.3 kWh m^{-2}) (www.ncbi.nlm.nih.gov). While solar energy is an obvious choice because it is abundant, (e.g. 1 h of sunshine, if harnessed properly, can sustain the whole globe for a year) renewable and leaves no carbon foot print. With the use of current technologies that can harness solar energy, it is possible to have small solar units that can concentrate solar energy for immediate and localized use for different purposes.

Solar collectors, concentrators and dryers in food processing

Dehydration is one of the most important step for preservation and value addition of food products through moisture control. Dehydrated foods have higher shelf life, making them available throughout the year even in off season. Because of their relatively low weight and low volume, dehydrated products are easy to store as well as to transport across distances. Different type of commodities which can be dried in the solar drying systems are mentioned in Table. 23.1

Table 23.1: Solar dried Agricultural commodities traded in market

Type	Examples of commodities
Food	Sugarcane, maize, wheat, rice, paddy, cow milk, vegetables, potatoes, cassava, sugar beet, soybeans, tomatoes, barley, meat, watermelons, bananas, onions, cocoa, coffee, palm oil, fruits etc.
Non food	Timber, wool, rubber, cotton etc.

Different types of solar dryers, collectors and concentrators can be used for various steps in food processing and value addition. Conduction, convection and radiation are the basic techniques by which water is forced to vaporise and the resulting vapour is removed either naturally or by force resulting in dehydration. Conventional convective drying is used for drying fruits and vegetables. However, this process also brings in some important changes in physical and chemical properties such as loss of color and change of texture, flavour and loss of nutrients.

The process of drying in the solar dryer is facilitated by circulation of hot air, spreading density of the product, the nature of pre-treatment as well as the nature of the product to be dried itself. The time taken for drying is also determined by the factors such as the initial moisture content and the desired final moisture content of the product. For example, it may take 4 h to reduce moisture content from an initial value of 4% to atmost nil for 50 kg refined wheat flour whereas 6 h are needed to reduce moisture content from an initial value value of 63 % to final level at 4% for 4 kg load of curry leaves (Eswara and Ramakrishnarao 2013).

While high temperature used during the processing are responsible for the alterations, lowering temperature increases the time of dehydration therefore increase in cost. Since fresh fruits and vegetables contain high water content (over 80 % moisture), the process of dehydration to a desirable lower moisture content such as 5–10 % (e.g. 75 kg of water from a 100 kg sample) is very energy-consuming. The process of dehydration alone contributes up to 30 % of the total cost of processing of most fresh produce. Thus, the cost of dehydration and energy consumption and quality of dried products play very important role in choosing an appropriate drying process.

There are various benefits from solar drying of the crop as it significantly improve the product quality in terms of colour, texture and taste in comparison to open sun drying. The crop is not contaminated by insect, pests and microorganisms.

Drying time is also reduced in many cases up to 50 per cent. Bulk density is reduced so storage of product is easy and it also increases the shelf life of the crop to be used in different seasons.

Solar dryers can be classified as direct or indirect and passive or active solar energy drying system. Passive solar energy drying systems are conventionally termed as natural circulation solar drying system and active solar energy drying systems mostly termed as hybrid solar dryers. The working principle of these dryers mainly depends upon the method of solar energy collection and its conversion to thermal energy for drying. In open sun drying, the crop is generally spread on ground, mat, cement floor where they receive short wavelength solar energy during a major part of the day and also natural circulation. A part of the energy is reflected back and remaining is absorbed by the surface of the crop depending upon the colour of the crop. The absorbed radiation is converted into thermal energy and the temperature of the material starts to increase. However there are losses like the long wavelength radiation loss from the surface of the crop and also convective heat loss due to the blowing wind through moist air over the crop surface. The process is independent of any other source of energy except sunlight and hence the cheapest method however, has a number of limitations as discussed. In general the open sun drying method does not fulfil the required quality standards of international market. With the adequate information of the problems of open sun drying a more scientific method of solar energy utilization for crop drying has emerged as solar drying systems active or passive systems more appropriately depending upon the method of solar energy collection and conversion to thermal energy.

Direct solar drying

The direct solar drying is also known as cabinet drying. In this method a part of the solar incident radiation on the glass cover is reflected back to the atmosphere and the remaining is transmitted inside the cabinet. A part of the transmitted radiation inside the cabinet then reflected back from the crop surface and rest is absorbed by the surface of the crop which causes its temperature to increase and thereby emit longwave radiations which are not allowed to escape in atmosphere due to the glass cover. The overall phenomenon causes the temperature rise of the crop. The glass cover in the cabinet dryer thus serve as a source for reducing the losses due to convection which plays an important role in increasing the crop and cabinet temperature.

Indirect solar drying

Indirect dryers contain vertical shelves to load the crop inside an opaque drying cabinet and a separate unit termed as solar collector is used for heating of the entering air. The heated air is allowed to flow through/over the wet crop that provide the heat for moisture evaporation by convective heat transfer between the hot air and the wet crop. Drying takes place due to the difference in the moisture concentration between the drying air and air in the vicinity of crop surface. The advantages of the indirect solar drying are that they offer good control of drying conditions and quicker, better quality produce is obtained as compared to open sun drying. It also helps in preventing localized heat damage from direct solar radiation due to opaque chamber. These dryers attain higher temperature than direct dryers thus recommended for deep layer drying. Due to prevention from direct radiation they are highly suitable for crops which are photo sensitive. However, they need more capital investment than direct dryers but it is worthwhile keeping in view the quality of produce and efficiency of indirect solar dryers.

Hybrid solar drying

This type of drying combines the features of both the direct and indirect type of solar drying. In hybrid type of dryers thermal energy is generated through the action of incident direct solar radiation and pre-heated air in a solar collector heater. Hybrid dryers consist of the same typical structure a solar air heater, separate drying chamber and a chimney and in addition has glazed walls inside the drying chamber (Sodha et al. 1987). Several type of material are used in such dryers to absorb heat energy like granite, mixture of waxes etc (Excell et al. 1980; Ayensu and Asiedu 1986).

Use of solar energy to prevent post-harvest losses

Being the second largest global horticultural producer after China, annual growth rate of fruit production in India is higher than vegetable production. To boost the production and productivity of vegetables and fruits in the country Government has been implementing Horticulture Mission for North East and Himalayan States(HMNEH) and National Horticulture Mission (NHM). The remaining states are covered under Mission for Integrated Development of Horticulture (MIDH). These schemes provide support for production of planting material, high yielding varieties of vegetable seed, rejuvenation of senile orchards, protected cultivation, creation of water resources, creation of infrastructure to prevent post-harvest losses of horticultural crops and for adoption of Integrated Nutrient Management (INM)/Integrated Pest Management (IPM)

Post-harvest losses

Recent studies indicate that while postconsumer food waste accounts for the greatest overall losses among affluent economies, food wastes are much higher at the immediate post-harvest stages in developing countries and wastage of perishable foods is higher acrossdeveloping economies. It has been estimated that 30–40 % of food is wasted in India every year for lack of a systematic post-harvest processing and preservation. India is losing ₹ 92,651 crores annually during post harvest phase (as per Post Harvest Loss Survey: 2007-12) which is about 3-times the annual budget for Indian agriculture.

Demand and supply gap of agricultural commodities

Demand and supply projections act as indicators to policy makers to formulate their medium and long-term agricultural policies. The present data shows that the increase in total demand is mainly due to growth in population and per capita income. The gap between supply and demand is narrowing down over the years for all the food items. The supply-demand gap for total cereals is expected to be 21.19 mt in 2011 whereas, it is projected at 16.96 mt in 2026 as shown in table 23.2 (Chand 2007).

Table 23.2: Supply-Demand gap for selected food items (Unit: million metric tons)

Food Items	Gap (Supply-Demand)		
	2011	2021	2026
Total Cereals	21.19	-2.94	-16.97
Pulses	-8.05	-24.92	-39.31
Edible oil	-6.66	-17.68	-26.99
Sugar	-4.31	-39.67	-74.13

Quality assessment of solar dried food products

The quality of solar dried products in terms of physical, chemical/nutritional and microbial properties is very important which can well establish its utility and appropriateness for crop drying. The extent of changes in each of these parameters during drying also depends on the pre-treatments which are the processes and steps taken during the preparation of the product for dehydration.

Physical parameters: Structure, case hardening, collapse, pore formation, cracking, rehydration properties, caking and stickiness are some of the physical parameters that are important for any kind of dried product. A number of factors influence these parameters during drying process. For example hot air drying destroys cell structure causing more time for rehydration of the product whereas freeze drying keeps the cell architecture almost intact. This is the most important attribute of freeze drying which highly affect the textural property and resultant affects on the dried product. However, it is always superior to any other kind of drying but it is highly expensive, sophisticated and energy consuming which is not always feasible. Pre-treatments such as immersion into alkaline or acidic solutions of citrates, carbonates have effect on colour while treatment with oleate esters or osmo-dehydration before drying create more porous structure therefore better rehydration property with only a minor effect on product quality.

Chemical/nutritional parameters: Fruits, vegetables and their products in the dried form are good sources of energy, minerals and vitamins. In the drying process there are certain changes in the food which affect its nutritional value such as loss of important vitamins and heat sensitive nutrients. Different type of chemical changes occur during drying of the food product such as enzymatic and non enzymatic browning, loss of vitamins and minerals, degradataion of coloring pigments, lipid oxidation and generation of resultant off flavours. Due to the reaction between reducing sugars and amino acids brown colour pigments are developed in food product which causes undesirable colour and flavour development and deterioration of food quality which effects its acceptance to consumer. Heating, sulphur dioxide or sulphites and acids are used to control enzymatic browning, though they are known to destroy vitamin B (thiamine). Dipping in citric acid/phosphoric acid/ascorbic acid solutions can inhibit enzymatic browning in fruits during osmosis. Heating causes Lipid oxidation which is responsible for rancidity, due to development of short chain fatty acids and trans-fatty acid and which are prone to oxidation and produces undesirable flavour and odour in the food, and loss of fat-soluble vitamins and pigments in many foods, especially in dehydrated foods. Many studies have shown that heating such as blanching before drying, boiling, steaming on drying itself causes degradation of important pigments such as α-carotene, β-carotene and lycopene. Factors that influence the oxidation rate of food product during drying and storage include moisture content, type of substrate (fatty acid), extent of reaction and oxygen content. Most fruit and vegetable products developed by solar dehydration/drying are pre-treated with proper antioxidants such as ascorbic acid in their formulations and improved packaging (under reduced oxygen),or vaccume packaging have been found to be helpful in minimizing rancidity and oxidation induced changes.

Vitamins such as A, C and thiamine are heat sensitive and sensitive to oxidative degradation. Sulphuring can destroy thiamine and riboflavin while pre-treatments such as blanching and dipping in sulphite solutions reduce the loss of vitamins during drying. As much as 80 % decrease in the carotene content of some vegetables may occur during open sun drying if they are dried without enzyme inactivation such as blanching. However, if the product is adequately blanched then carotene loss can be reduced to 5 %. Steam blanching retains higher amounts of vitamin C in spinach compared with hot-water blanching (Ramesh et al. 2001). Blanching in sulphite solution can retain more ascorbic acid in okra (Inyang and Ike 1998). Sodium metabisulphite treatment was able to reduce oxidation of carotenoid in carrots and L-cysteine-HCl help in retaining highest amount of ascorbic acid (Mohamed and Hussein 1994).

Microbial parameters: Moisture content highly affect the growth of microorganisms in food. When water activity is high the chances of overall microbial load will be higher. Thus, by controlling moisture level and pH, microbial growth can be controlled but does not result in a sterile product. Drying can significantly reduce the overall microbial load due to two reasons firstly it decreases the water activity so microorganisms cannot grow in food due to unavailability of required moisture and secondly due to application of heat, microbial load present in food is significantly reduced. Similarly blanching can reduce microbial load but cannot eliminate spore forming organisms. The type of microflora present in dried products depends on the characteristics of the dried products, such as pH, composition, pre-treatments, types of endogenous and contaminated microflora and method of drying. Solar dried fruit and vegetable products are routinely tested for microbial load and are found to be within the allowed safe limits. This is also the reason behind the long shelf life for these products.

Conclusion

Solar energy based technologies can play an important role in food processing. However, at present their use is very limited and generally used in to micro- to small-scale level processing. To make the solar food processing technology applicable for large scale processing is a challenge and its versatility is another issue. Solar dehydration/drying is as effective as the mechanized equipments and offers an alternative at low or no cost. However, solar drying is highly dependent on weather conditions, thus alternative arrangements for continued drying is ery important to make it sustainable for large scale processing. A limited number of governmental incentives that are currently in place through various ministries (e.g. Ministry of New and Renewable Energy, India) help fray the initial high cost of the solar gadgets and help reduce the pay-back time for small and micro-entrepreneurs but in case of application of solar energy in food-processing industry has not progressed as fast. There is a need for integration of food technologists and renewable energy scientists to develop the solar food processing eqquipments as per the need of the processing industry requirements. There is a huge growth possibility for food processing sector in the Indian market and there is also a feasibility for application of solar gadgets in processing sector due to the geographical positioning of the country as in major parts tropical conditions prevails with sunny weather conditions. There is a need to work on the policy level for promotion of solar appliances on the industrial level from the government side also to foster the use of renewable energy in this high energy consuming sector. Techno-economic awareness of the solar gadgets with high flexibility to work in the adverse weather conditions is very important to get fitted in the food processing industry where the operations are continuous and of interdependent nature. The cost is also a major determining factor which should be in the reach of the small micro finance units.

A combination of solar gadgets rather than any single one will ultimately take care of major energy needs in food processing industry. This is already happening with considerable success. As the promise of reduced cost of photovoltaic and collectors is realized and as newer technologies become reliable and affordable, solar power generators may produce most of the world's electricity in the next 50 years, photovoltaic technologies and concentrated solar power together can become the major source of electricity dramatically as well as reducing the emissions of greenhouse gases that harm the environment. Solar power will certainly be a beneficiary as well as a contributor in the future of the food processing sector in India and the world.

References

Ayensu, A., Asiedu-Bondzie, V., 1986, Solar drying with convective self-flow and energy storage, Solar and Wind Technology, 3(4), pp. 273-279.

Chand, R. 2007. Demand of foodgrains. Economic and political weekly. (December) 10-13.

Eswara, A.R., Ramakrishnarao, M,2013, Solar energy in food processing—a critical appraisal. Journal of Food Science and Technology. doi: 10.1007/s13197-012-0739-3

Exell, R. H. B., 1980a, A basic design theory for a simple solar rice dryer, Renew Energy Rev J, 1(2), pp. 1-14.

Exell, R. H. B., 1980b, A simple solar rice dryer- basic design theory, Sunworld, 4(6), 186-190.

Inyang UE, Ike CI. Effect of blanching, dehydration method, temperature on the ascorbic acid, color, sliminess and other constituents of okra fruit. Int J Food Sci Nutr. 1998;49:125–130. doi: 10.3109/09637489809089392. [PubMed] [Cross Ref]

Jain, D., Tiwari, G. N., 2003, Thermal aspects of open sun drying of various crops, Energy 28, Jain, D., 2005a, Modelling the system performance of multi-tray crop drying using an inclined multi-pass solar air heater with in-built thermal storage, Journal of Food Engineering, 71 (1), pp. 44–54.

Jain, D., 2005b, Modeling the performance of greenhouse with packed bed thermal storage on crop drying applications, Journal of Food Engineering, 71 (2), pp. 170–178.

Jain, D., Pathare, P. B., 2007, Study the drying kinetics of open sun drying of fish, Journal of Food Engineering, 78 (4), pp. 1315-1319.

Hughes, B. R., Oates, M., 2011, Performance investigation of a passive solar-assisted kiln in the United Kingdom, Solar Energy, 85, pp. 1488-1498.
Majumdar, A.S. 2015, Handbook of Industrial Drying. CRC Press. 305pp.

Mekhilefa, S., Saidurb, R., Safari, A., 2011, A review on solar energy use in industries Renewable and Sustainable Energy Reviews, 15, pp. 1777–1790.

Mohamed S, Hussein R. Effect of low temperature blanching, cysteine-HCl, N-acetyl-L-cysteine, Na metabisulphite and drying temperatures on the firmness and nutrient content of dried carrots. J Food Process Preserv. 1994;18:343–348. doi: 10.1111/j.1745-4549.1994. tb00257.x. [Cross Ref]

Ramesh MN, Wolf Tevini D, Jung G. Influence of processing parameters on the drying of spice paprika. J Food Eng. 2001;49:63–72. doi: 10.1016/S0260-8774(00)00185-0.[Cross Ref]

Sodha, M. S., Bansal, N. K., Kumar, K., Bansal, P. K., Malik, M. A. S., 1987, Solar Crop Drying 1, West Palm Beach, CRC Press.

Visavale, G. L., Sutar, P. P. and B. N. Thorat., 2011, Comparative Study on Drying of Selected Marine Products: Bombay Duck (Herpodon nehereus) and Prawn (Penaeus indicus), International Journal of Food Engineering, Vol. 7: Iss. 4, Article 20.

Visavale, G.L., 2009, Design and Characteristics of Industrial Drying Systems. Ph. D. thesis, Institute of Chemical Technology, Mumbai, India.

Xingxing, Z., Xudong, Z., Stefan, S., Jihuan, X., Xiaotong, Y., 2012, Review of R&D progress and practical application of the solar photovoltaic/thermal (PV/T) technologies, Renewable and Sustainable Energy Reviews, 16, pp. 599-617.

FAOSTAT Website (http://faostat3.fao.org/home/E) as on 1st July 2015.

https://www.ncbi.nlm.nih.gov/pmc/articles/PMC3550910/

Printed in the United States
by Baker & Taylor Publisher Services